RAND NATIONAL SECURITY RESEARCH DIVISION

Defense Resource Planning Under Uncertainty

An Application of Robust Decision Making to Munitions Mix Planning

Robert J. Lempert, Drake Warren, Ryan Henry, Robert W. Button,
Jonathan Klenk, Kate Giglio

Prepared for the Office of the Secretary of Defense

For more information on this publication, visit www.rand.org/t/RR1112

Library of Congress Control Number: 2016931459

ISBN: 978-0-8330-9167-3

Published by the RAND Corporation, Santa Monica, Calif.

© Copyright 2016 RAND Corporation

RAND® is a registered trademark.

Support RAND

Make a tax-deductible charitable contribution at
www.rand.org/giving/contribute

www.rand.org

Preface

The goal of this project is to help improve the value and character of defense resource planning in an era of growing uncertainty and complex strategic challenges. Because it is impossible to predict what threats may arise and how defense funding will progress, a new approach is needed to develop robust resource strategies, that is, strategies that perform well over a wide range of threat and funding futures and thus are better at managing surprise.

To address this need, RAND researchers applied a proven approach to strategy discovery, Robust Decision Making, or RDM, to defense planning. RDM, a quantitative decision support methodology for informing decisions under conditions of deep uncertainty and complexity, has been applied to many policy areas in the last decade. This document explores how the RDM method may be applied to defense resource planning in an application to air-delivered conventional munitions mix planning.

This research was sponsored by the Cost Assessment and Program Evaluation (CAPE) Directorate within the Office of the Secretary of Defense (OSD) and conducted within the International Security and Defense Policy Center of the RAND National Defense Research Institute, a federally funded research and development center sponsored by the Office of the Secretary of Defense, the Joint Staff, the Unified Combatant Commands, the Navy, the Marine Corps, the defense agencies, and the defense Intelligence Community.

For more information on the RAND International Security and Defense Policy Center, see www.rand.org/nsrd/ndri/centers/isdp or contact the director (contact information is provided on the webpage).

Contents

Figures

Tables

Summary

This study applies Robust Decision Making (RDM)—an approach to management under conditions of deep uncertainty—to the challenge of defense resource planning. Defense planning faces many difficult and conflicting requirements. It must allow detailed comparisons among many complicated options, so that the nation can reliably and cost-effectively meet its military needs. Defense planning needs to enable coordination among large, complicated organizations. It needs to guide investments, some of which have years-long lead times. It is expected to provide transparency and accountability to the public. And it needs to recognize that most long-range predictions are wrong and that the future is sure to surprise.

Concerned that the current approaches to defense planning rest too heavily on assumptions that may not hold, the Cost Assessment and Program Evaluation (CAPE) Directorate within the Office of the Secretary of Defense (OSD) asked RAND to evaluate the utility of RDM. Although its distant origins lie in defense planning, RDM has more recently matured and has seen widespread application in the areas of energy, environment, climate, infrastructure, and insurance. As an initial test case, CAPE asked RAND to apply RDM and evaluate its utility for one of the archetypal challenges of defense planning under uncertainty: the munitions mix problem. This report addresses in detail how RDM might be applied to this specific problem, but the reader should bear in mind that the purpose of the report is not to present a recommended solution for managing the munitions mix but rather to investigate the applicability of RDM to a broad set of defense challenges involving deep uncertainty.

Planning with Predictive Failure

Traditionally, the Department of Defense (DoD) has conducted defense resource planning by employing the *predict-then-act* approach. Analysts assemble available evidence into best-estimate predictions of the future and then use models and tools to suggest the optimal strategy given these predictions. Such methods work well when the predictions are accurate and not controversial. Otherwise, the methods can produce gridlock and lead to solutions that fail when the future turns out differently than expected.

Since the end of the Cold War, DoD has been operating in a security environment that has become one of surprise and uncertainty and is virtually certain to confound even the best predictions. Although DoD has taken steps to improve the range of its predictions, there is no guarantee that such improvements will keep pace with the inherent unpredictability of today's long-term national security challenges. Many defense analysts now argue that DoD plans should assume a high likelihood of predictive failure of any postulated future (e.g., Danzig, 2011).

RDM is an iterative, quantitative, decision support methodology designed to address the challenges of predictive failure. The approach has been applied to areas outside national security, such as flood risk (Fischbach, 2010; Lempert et al., 2013a) and water management applications (Groves and Lempert, 2007; Groves et al., 2008; Means et al., 2010) in situations where decision makers face conditions of deep uncertainty. Deep uncertainty occurs when the parties to a decision do not know—or agree on—the best model for relating actions to consequences or the likelihood of future events (Lempert, Popper, and Bankes, 2003).

RDM rests on a simple concept. Rather than using models, data, and constraining assumptions to describe a best-estimate future, RDM runs models using hundreds, thousands, or even millions of different sets of assumptions to describe how plans perform in many plausible futures. The approach then uses statistics and visualizations on the resulting large database of model runs to help decision makers identify those future conditions where their plans will perform well and poorly. This information can help decision makers develop plans that are more robust to a wide range of future conditions.

In contrast to the predict-then-act approach, RDM runs the analysis "backward," using a *vulnerability-and-response* approach. Analysts begin with one or more strategies under consideration (often a current plan) and then, using potentially the same models and tools, characterize a spectrum of future conditions, including some where a strategy fails to meet its goals (is vulnerable). This serves as a stress-test of strategies and helps decision makers identify "robust" strategies—those that perform reasonably well regardless of what the future brings—and identify the key tradeoffs among potential robust strategies. Often, the robust strategies identified by RDM are adaptive, designed to evolve over time in response to new information (Lempert, Popper, and Bankes, 2003).

The predict-then-act approach condenses information about a range of potential futures into a single best-estimate future (sometimes expressed as a probabilistic forecast) or a small number of planning scenarios. But RDM assembles the results of many hundreds, thousands, or even millions of computer simulation model runs and uses this database of runs to comprehensively explore and summarize the challenges and opportunities the future might bring. In particular, RDM provides a means to effectively communicate the information in these many runs by summarizing them as a small number of decision-relevant scenarios. By embracing many plausible futures, RDM can help reduce overconfidence and the deleterious effects of surprise, can systematically include imprecise information in the analysis, and can help decision makers and stakeholders with differing expectations about the future nonetheless reach consensus on action (Lempert and Popper, 2005; Groves and Lempert, 2007; Hallegatte et al., 2012).

Munitions Mix Challenge

CAPE and RAND agreed to evaluate RDM by applying it to the air-delivered conventional munitions mix challenge. DoD must purchase sufficient weapons for its platforms to enable U.S. forces to complete their missions. Deciding which munitions to purchase is complicated because future conflicts and associated missions are uncertain, and many different types of weapons exist, each serving different purposes and carried by different platforms. Some weapons are very expensive but others are less so; likewise, some are very sophisticated and some more rudimentary. Each campaign will typically require the use of specific weapons in particular ways, orchestrated across the entire campaign. For instance, many campaigns begin by using a relatively small number of expensive, standoff, precision-guided munitions (PGMs) to disable the adversary's air defenses. Subsequently, more numerous and less-expensive munitions destroy other targets.

Currently, DoD uses the Munitions Requirement Process (MRP) to inform its munitions purchase decisions. Each year, DoD and the services conduct the MRP to generate a total munitions requirement. The MRP first identifies a small number of specific scenarios called *illustrative planning scenarios* and then develops detailed target lists for each scenario. Each service—the Army, Navy, Air Force, and Marine Corps—then recommends its optimal weapons mix contingent on the planning scenarios and all the assumptions they contain.

The current MRP aims to avoid predictive failure in several ways. First, it includes a small number of alternative planning scenarios, whose goal is to approximate the range of relevant future conditions that DoD's munitions mix will need to address. A wealth of detail is added to each scenario to further reduce uncertainty. These details include specific limiting assumptions about global and regional security conditions, local operational conditions, various participants' alignment and force postures, and choices by a range of actors involving priorities, operations, and risk profiles. Finally, safety factors in the form of increased weapons requirements are added to compensate for those uncertainties not considered in the analysis.

The MRP is often criticized for overstating requirements and for purchasing inadequate quantities of some munitions. Further, the MRP provides little information to DoD regarding the security environments or specific contingencies for which its weapons mix is likely to prove sufficient and those for which it is likely to fall short. These shortcomings seem inherent in the predict-then-act approach that underlies the MRP, along with many other areas of defense planning. Each illustrative planning scenario contains hundreds to thousands of assumptions. These constraining assumptions decrease the flexibility of the analysis and thus its ability to explore over a wide range of security environments and contingencies. This limits the opportunities for participants in the process to tune assumptions so that the analysis generates robust policy recommendations.

A More Robust Munitions Mix Strategy

This study conducted two iterations of an RDM analysis. The first iteration focused on a broad array of munitions mix strategies developed to address relatively simple planning scenarios. The second iteration included strategies that were based on the findings of the first but focused on a narrower range specifically designed to be robust over a wide range of futures.

CAPE staff participated actively in both iterations of the analysis, helping to define the munitions mix strategies considered, the uncertainties they face, and the metrics and models used to evaluate their success.

Many uncertain factors may affect the success of a munitions mix strategy. In particular, this study aims to examine potential interplays between large- and small-scale factors. Traditional defense planning scenarios are generally differentiated by large-scale factors, such as the size of the campaign (small or large), where it is fought (e.g., Asia, the Middle East), and the military nature of the enemy (developing country or near-peer). But each scenario also contains numerous small-scale assumptions, such as the effectiveness of particular weapons against particular targets or the resilience of the enemy. The extent to which policy implications drawn from such scenarios depend on the large- or small-scale factors, or some combination, may not be clear. This study thus considers many futures that explore alternative combinations of large- and small-scale factors. In particular, we characterized potential future security environments by assembling a list of the 35 conflicts fought by the United States over the last century and grouped these conflicts into six categories of increasing intensity. We then constructed 25 alternative future security environments, each a 20-year window (covering most munitions' life cycle) of different combinations of conflicts in these six security environments, as shown in Figure S.1. The first of these security environments represents a repeat of the last 20 years. The other 24 were chosen to provide a diverse range of environments. For each security environment, we considered 50 randomly chosen combinations of parameters representing the small-scale factors, for a total of 1,250 futures against which to stress-test the munitions strategies.

We built two coupled models for this study to link policy choices to outcomes. The weapons on target (WoT) model uses weapons inventories to fight individual campaigns. The campaign generator (CG) generates 20-year sequences of campaigns and provides the munitions to fight them. The WoT model simulates individual campaigns on a day-to-day basis, using an optimization algorithm to match munitions and delivery vehicles to targets. WoT is similar to such campaign models as the Air Force's Combat Forces Assessment Model and related models used by CAPE, but with less detail. The CG provides a series of campaigns and their attributes

Figure S.1
Future Security Environments Used in the RDM Analysis

	SE	2014	2016	2018	2020	2022	2024	2026	2028	2030	2032
Last 20 years	21	A	A	B	D	E	D	D	D	C	B
Benign	1	A	A	B	B	B	A	A	A	A	A
Moderate to benign	23	C	C	C	C	A	A	A	A	A	A
Benign to challenging	24	A	A	A	A	A	B	B	B	D	D
Benign+	2	A	B	B	B	E	A	A	C	C	A
Challenging to benign	22	E	E	C	C	A	A	A	A	A	A
Moderate	20	E	B	B	B	A	A	B	B	B	E
Volatile	25	A	E	A	E	E	A	E	A	A	A
Moderate (steady)	4	B	B	B	D	D	B	B	B	D	D
Moderate (spikes)	3	A	E	B	B	B	E	A	C	C	E
Moderate	17	C	C	F	F	A	B	B	B	B	B
Benign to challenging	19	A	B	B	B	B	B	B	F	F	D
Difficult (steady)	5	E	D	D	C	C	B	B	B	C	C
Difficult (surge)	6	C	C	F	F	B	B	B	A	A	E
Difficult (surge) 2	8	C	C	B	B	B	F	F	A	A	E
Difficult	14	E	B	B	B	D	D	E	A	C	C
Difficult (increasing	13	B	B	B	B	B	B	F	F	D	D
Moderate	16	C	C	B	B	B	E	C	C	E	D
Difficult (varied)	10	D	D	A	E	A	F	F	B	B	B
Difficult (varied, increasing)	12	A	A	A	F	F	E	A	D	D	D
Challenging +	7	F	F	C	C	B	B	B	C	C	E
Challenging + 2	9	C	C	F	F	B	B	B	C	C	E
Challenging + 3	11	D	D	A	A	F	G	E	C	C	B
Challenging (increasing)	15	B	B	B	F	F	D	D	A	D	D
Challenging (increasing)	18	E	B	B	B	A	D	D	F	F	D

NOTES: Security environments (SEs) are ordered so that the last 20 years are listed first, and then entries are ordered according to increasing severity. SE A is the most benign and SE F is the most stressful.
RAND RR1112-S.1

to the WoT model. The two models are configured so that we can run thousands of alternative futures for each of several munitions mix strategies.

As with many policies, munitions mixes are adaptive, adjusting over time in response to new information. Much less often are policies designed to be adaptive. A strategy so designed takes near-term actions with explicit consideration of how they might be subsequently adjusted. A strategy designed to be adaptive may also include systematic consideration of how it will gather information and respond to it in the future. Traditionally, analyses of munitions mix strategies have not considered how such mixes might adjust over time. Rather, they focus on describing some single best mix at a single point in time. This study takes a first step toward considering adaptive munitions mix planning strategies. We characterize each strategy with two components: *desired portfolio goals* that specify the number of each type of munition that policymakers would like to have in the U.S. stockpile and *purchasing rules* that describe how munitions will be purchased to replenish the stockpile when it is depleted during one or more campaigns. The portfolio goals are related to the alternative force sizing constructs often used in DoD planning. We find consideration of purchasing rules important. In many futures in our analysis, munitions mix strategies fail because they do not appropriately restock inventories between closely spaced campaigns.

To compare the performance of alternative munitions purchasing strategies in each of the many futures, this RDM analysis uses two measures: cost and military sufficiency (i.e., campaign success rates and the amount of time to complete successful campaigns).

The findings of this analysis emerge from simplified models and unclassified data, so at best are suggestive rather than definitive. Nonetheless, this analysis finds that a munitions mix strategy that we call *Big+Deter-Mixed* is robust over a wide range of plausible futures. The *Big+Deter-Mixed* strategy consists of a portfolio goal, which specifies the desired number of weapons of each type, and a purchase rule, which specifies the order in which munitions will be replaced when the stockpile is depleted. We constructed the portfolio goal by using a WoT optimization model to give the weapons mix that provides the best balance between weapons' acquisition cost and time to completion for two planning scenarios: (1) a deterrence campaign with a small number of targets accessible only to standoff weapons and (2) two back-to-back medium-size campaigns. We chose this set of planning scenarios through an iterative process of stress-testing strategies with portfolio goals derived from alternative sets of planning scenarios. *Big+Deter-Mixed* uses a purchase rule we call replenishment, which restocks weapons inventories in proportion to shortages in the inventories. We considered two alternative purchase rules and chose replenishment as the superior one.

We stress-tested *Big+Deter-Mixed* and five alternative strategies over a wide range of futures that combine assumptions about both large-scale factors—alternative security environments with varying levels of severity—and small-scale factors—alternative values for parameters representing weapons effectiveness, adversary's capabilities, and tactical decisions. We evaluated the strategies using three measures of performance: their ability to complete campaigns (success rate), the speed with which they complete campaigns (days to completion), and the total cost of acquiring and replenishing the weapons portfolio (cost).

A scenario discovery statistical cluster analysis, applied to the model-generated database of thousands of futures, identifies two important scenarios around which we organize our comparisons of the strategies. The Moderate Scenario contains those futures in which *Big+Deter-Mixed* has a generally high success rate (greater than 90 percent) and the Extreme Scenario contains those futures where *Big+Deter-Mixed* has a generally low success rate (less than

90 percent). The most important uncertainties distinguishing these two scenarios are the severity of the security environment and the effectiveness of Global Positioning System (GPS) weapons. These two scenarios—Moderate and Extreme—suggest that with a high level of GPS effectiveness and in security environments up to the severity level of the last 20 years, *Big+Deter-Mixed* will have a generally high success rate over a wide range of assumptions about other uncertainties. If GPS effectiveness is low, the *Big+Deter-Mixed* strategy will have high success rates only in security environments about half as severe as those in the past 20 years. The scenarios produced by this analysis helped organize the rest of the study and also demonstrate how in RDM analyses the planning scenarios are *outputs of* rather than *inputs to* the analysis.

Of the six strategies considered in this analysis, *Big+Deter-Mixed* is the most robust in both scenarios, in the sense that it performs better than the alternatives for each of the three measures—success rate, completion time, and cost—over a wide range of futures.

Figure S.2 summarizes the strengths and limitations of *Big+Deter-Mixed* over the wide range of futures in the Moderate Scenario. Each dot in the figure represents the performance of the *Big+Deter-Mixed* strategy in one future. The horizontal axis shows the days to comple-

Figure S.2
Days to Completion Regret, Success Regret, and Success Rates for the *Big+Deter-Mixed* Strategy for Each of the 361 Futures in the Moderate Scenario

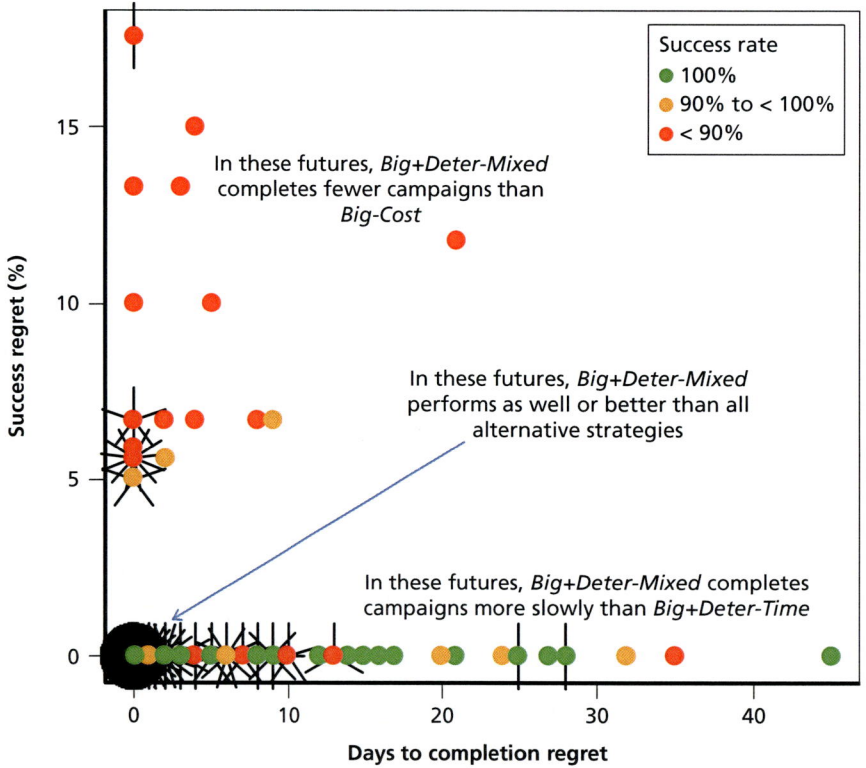

NOTES: Days to completion regret is calculated by comparing *Big+Deter-Mixed* days to completion against the days to completion of other strategies with the same or better success rate. For overlapping points, the whiskers show the number of overlapping points. No regret cases shows the distribution of points at the origin (zero days to completion regret and zero success regret).
RAND RR1112-S.2

tion regret, that is, how much longer *Big+Deter-Mixed* takes to complete its campaigns than the strategy in the future that completes them the fastest. The vertical axis shows the success rate regret, that is, how much lower *Big+Deter-Mixed*'s success rate is than the strategy with the highest success rate in that future. The color of each dot shows *Big+Deter-Mixed*'s success rate (without comparison to any other strategy). The black "whiskers" indicate the number of overlapping dots.

Figure S.2 shows some important patterns. First, *Big+Deter-Mixed* has zero regret in most futures in the Moderate Scenario. In 208 of these 361 futures, *Big+Deter-Mixed* performs as least as well or better than the other five strategies in both success rate and time to completion. This is the basis of our finding that *Big+Deter-Mixed* is the most robust munitions mix strategy considered in this study.

However, in a small number of futures, *Big+Deter-Mixed* does not perform as well as the alternatives. In the futures along the horizontal axis, a strategy we call *Big+Deter-Time* completes campaigns faster than *Big+Deter-Mixed*. This makes sense, because the former strategy has a portfolio optimized to complete campaigns quickly. But although slower, *Big+Deter-Mixed*'s success rate is at least as good as that of *Big+Deter-Time* in these futures. In the futures along the vertical axis, however, a strategy we call *Big-Cost* has higher success rates than *Big+Deter-Mixed*. This is surprising; *Big-Cost* has a portfolio optimized to save money, so that the strategy costs about a fourth as much as *Big+Deter-Mixed*.

Analyzing in detail the futures along the vertical axis helps explain this vulnerability of the *Big+Deter-Mixed* strategy. In futures with closely spaced conflicts, the strategy spends its resources purchasing expensive PGMs after the first conflict and has not sufficiently replenished its stockpiles when the second campaign begins. In contrast, *Big-Cost* purchases a large number of economical PGMs after the first campaign and is more quickly ready for the second.

This RDM vulnerability analysis usefully suggests ways to adjust *Big+Deter-Mixed*'s purchasing rule that might eliminate this vulnerability with respect to the *Big-Cost* strategy. In particular, it suggests that a step-wise purchase rule that first invests toward the *Big-Cost* portfolio and—once achieved—switches toward purchasing more expensive weapons might eliminate many of the vulnerabilities of the *Big+Deter-Mixed* strategy.

In the Extreme Scenario, *Big+Deter-Mixed* generally performs better than the alternative strategies but has insufficiently high success rates, because the scenario's campaigns require far more weapons than can be purchased. To reduce these vulnerabilities, the United States could spend more on munitions or consider other policy options, such as developing a surge capability in the munitions industrial base or ensuring sufficient warning time to increase munitions spending well in advance of large campaigns.

The Future of RDM in Defense Planning

This initial application of RDM to the munitions mix challenge provides a proof of concept showing how RDM might be employed for defense planning. This application demonstrates that RDM can provide the types of analytic information DoD might find useful in identifying robust and flexible strategies that can achieve success despite predictive failure.

Future RDM defense planning applications would need to address a number of challenges. First, they might need to draw on more powerful computational resources than were available for this study. The analysis here used relatively simple simulation models with no classified data but still faced computational constraints. In particular, using complex models, it could take several weeks to conduct each iteration of the analysis, which reduces the rate at which we could design and explore new strategies. DoD and RAND have access to more capable computational platforms, such as the defense computing resources (e.g., DoD's High Performance Computing Modernization Program). But as RDM is integrated into defense planning, the computational requirements are likely to also increase as the models used become more complex.

Second, future RDM applications might need to employ a wider set of models, and ones not of the type currently used in formal DoD planning exercises. For instance, the importance of purchasing rules in the Moderate Scenario and the demands of the Extreme Scenario suggest that in seeking robust strategies, an RDM munitions mix analysis might usefully consider policies that affect the munitions industrial base and logistics system, in addition to WoT campaign models. Traditionally, the MRP employs only campaign models, so a more complete RDM analysis might call for a significant expansion of the elements of the full system considered in DoD's munitions mix studies.

Finally, integrating RDM into defense planning could raise a number of organizational challenges. Initially at least, and perhaps as a permanent approach, RDM might be used as a precursor to or in parallel with traditional processes. For instance, RDM could stress-test munitions mix strategies generated by the current MRP over a much wider range of futures and suggest new scenarios in which these strategies might be considered. In addition, RDM could provide guidelines for potential robust strategies, which could then be tested and fleshed out using traditional tools for defense planning. Alternatively, RDM analyses with very simple models might serve as useful screening tools to situate more traditional, detailed analyses. Future research and experience might help clarify the situations in which RDM analyses might use simpler or more complicated models as part of overall, integrated planning processes.

Above and beyond these challenges of conducting and integrating different types of analysis, RDM might raise challenges of process and communication for DoD. For instance, RDM often stress-tests strategies until they break. This provides useful information, but DoD might have to employ security-sensitive procedures to internally and externally manage and communicate information regarding the vulnerabilities of their proposed policies.

We recommend that RDM be used to initially supplement current defense planning. As RDM continues to validate itself in the national security analytical realm, as it has in infrastructure planning, it can be better integrated into the planning, programming, and budgeting system and deliberative planning activities. Shifting the defense planning processes to a more balanced approach, through the integration of RDM, would involve issues of adjustment, but we expect that RDM's benefits will far exceed its costs—to the benefit of the Pentagon, the Congress, and the nation. RDM would improve defense planning by enabling DoD to examine its strategies, policies, plans, programs, and budgets over a wide range of futures; identify vulnerabilities; and design responses that reduce those vulnerabilities. This in turn would improve DoD's ability to design and evaluate robust and flexible strategies. RDM can help DoD more successfully achieve its goals in a world in which surprise and uncertainty are virtually certain to lead to predictive failure.

Acknowledgments

We are grateful to Web Ewell and Kay Sullivan in the Cost Assessment and Program Evaluation Directorate, Office of the Secretary of Defense, for their mentorship in conducting this research project. We thank our two reviewers, Preston Dunlap and Joel Predd, for their exceptionally thoughtful and helpful reviews.

Abbreviations

AARGM	Advanced Anti-Radiation Guided Missile
ATACMS	Army Tactical Missile System
BDA	bomb damage assessment
C2	command and control
CALCM	Conventional Air-Launched Cruise Missile
CAPE	Cost Assessment and Program Evaluation
CG	campaign generator
COIN	counterinsurgency operations
DMPI	designated mean point of impact
DoD	Department of Defense
DR	delivery rate
ER	extended range
GAM	Global Positioning System Guided Munition
GDP	gross domestic product
GPS	Global Positioning System
GWoT	global war on terror
HARM	High-Speed Anti-Radiation Missile
JASSM	Joint Air-to-Surface Standoff Missile
JDAM	Joint Direct Attack Munition
JSOW	Joint Standoff Weapon
MF	munitions funding
MOP	massive ordnance penetrator
MRP	Munitions Requirements Process

OpTempo	operations tempo
OSD	Office of the Secretary of Defense
PGM	precision-guided munition
Pk	probability of kill
POL	petroleum, oil, lubricants
PPBS	planning, programming, and budgeting system
RDM	Robust Decision Making
RF	radio frequency
SAM	surface-to-air missile
SDB	small diameter bomb
SE	security environment
SLAM	Standoff Land-Attack Missile
TLAM	Tomahawk Land Attack Missile
WMD	weapons of mass destruction
WoT	weapons on target
XLRM	uncertainties, policy levers, relationships, measures

CHAPTER ONE

Introduction

This study aims to apply Robust Decision Making (RDM)—an approach to management under conditions of deep uncertainty—to the challenge of defense resource planning. Defense planning faces many difficult and conflicting requirements. It must allow detailed comparison among many complicated options, so that the nation can reliably and cost-effectively meet its military needs. Defense planning must enable coordination among large, complicated organizations. It must guide investments, some of which have years-long lead times. It must provide transparency and accountability to the public. And it must recognize that most long-range predictions are wrong and that the future is sure to surprise.

Concerned that the current approaches to defense planning rest too heavily on assumptions that may not hold, the Cost Assessment and Program Evaluation (CAPE) Directorate within the Office of the Secretary of Defense (OSD), asked RAND evaluate the utility of RDM. Although its origins lie in defense planning,[1] RDM has more recently matured and has seen widespread application in the areas of energy, environment, infrastructure, and insurance. As an initial test case, CAPE asked RAND to apply RDM and evaluate its utility for one of the archetypal challenges of defense planning under uncertainty: the munitions mix problem.

Planning with Predictive Failure

In a recent essay, Richard Danzig, Former Secretary of the Navy, faults the Department of Defense (DoD) for an overreliance on prediction in its planning (Danzig, 2011). He notes that "the propensity to make predictions—and to act on the basis of predictions—is inherently human" and that this propensity "is especially deeply embedded at the highest levels of DOD" owing in part to the long and relatively successful Cold War competition against a reasonably predictable adversary. But today's security environment has become one of "surprise and uncertainty" that is virtually certain to confound any predictions. While lauding attempts to improve predictive accuracy, Danzig warns that there is no guarantee that such improvements will keep pace with the inherent unpredictability of today's long-term national security challenges. Rather, he argues, DoD plans should assume a high likelihood of predictive failure.

Fortunately, Danzig notes several types of practical strategies that DoD can pursue to achieve its goals even when its predictions prove wrong. DoD already employs many such strategies to some degree, as do planners in many other fields. Danzig suggests that DoD could

---

[1]   See, for instance, Bankes (1993) and Dewar (2002).

accelerate the tempo of its decision making processes, while delaying some decisions. It could increase the agility of production processes, prioritize investments in the equipment that is most adaptable, build more for the short term, and nurture diversity and competition. In brief, Danzig argues that DoD should plan to succeed in the face of predictive failure by emphasizing flexibility and robustness in its investments and planning processes.[2]

Achieving such robustness and flexibility may require that DoD incorporate new methods and tools for planning. RDM may provide one such approach. An iterative, quantitative decision support methodology, RDM is designed to help decision makers identify and compare robust strategies—ones that perform better than the alternatives over a wide range of plausible futures (Lempert, Popper, and Bankes, 2003; Lempert et al., 2006). The approach begins with one or more strategies under consideration. The analysis then runs a simulation model over many (hundreds to millions) of plausible paths into the future to create a database of model results. Visualization and cluster analysis tools then succinctly summarize those future conditions that best distinguish those in which the strategy meets its goals from those in which it does not. The analysis then helps decision makers use this information to identify potential responses to any vulnerabilities and the tradeoffs among them. The RDM process can reduce overconfidence by challenging analysts and decision makers to explore a wide range of plausible futures. RDM's design facilitates stakeholder deliberation and consensus by providing an analytic framework in which parties can agree on near-term actions without agreeing on long-term expectations and values. As one important attribute, RDM often helps decision makers to design adaptive (flexible) strategies, designed to evolve over time in response to new information, whose components may not be obvious at the onset.

RDM appears to have many attributes that could enable DoD to better implement Danzig's call to reduce risks from predictive failure by adopting flexible and robust strategies. In the past decade, RDM has increasingly served this role for organizations concerned with energy and environmental management. But as Danzig notes, defense planning has its own unique attributes and challenges. This study thus examines how RDM might fare in this context.

Munitions Mix Challenge

CAPE and RAND, after reviewing several options,[3] agreed to evaluate RDM by applying it to the air-launched munitions mix challenge. DoD must purchase sufficient weapons for its aircraft to enable U.S. forces to complete their missions. Deciding which munitions to purchase is complicated because future missions are uncertain, and many different types of weapons exist, each serving different purposes and carried by different platforms. Some weapons are very expensive, but others are less so. Each campaign will typically require the use of specific weapons in a particular sequence. For instance, many campaigns begin by using a relatively small

---

[2]   Many others have criticized defense planning in the United States. For example, Davis and Khalilzad (1996) said that "Traditional 'threat-based planning". . . is no longer an adequate basis for mid- and long-range planning" in the post–Cold War world and that defense planning should "confront head-on the reality of substantial uncertainty in many dimensions." More generally, Tetlock (2006) found that experts' predictions across a variety of issues were only slightly better than chance.

[3]   CAPE and RAND also considered using RDM to examine solutions to the intelligence, surveillance, and reconnaissance (ISR) force mix investment problem and the weapons of mass destruction (WMD) failed state problem.

number of expensive, standoff, precision-guided munitions (PGMs) to disable the adversary's air defenses. Subsequently, more numerous and less-expensive munitions destroy other targets.

Currently, DoD uses the Munitions Requirements Process (MRP) to inform its munitions purchase decisions.[4] Each year, DoD and the services conduct the MRP to generate a total munitions requirement. The MRP first identifies a small number of specific scenarios called *illustrative planning scenarios* and then develops detailed target lists for each scenario. Each target list is divided among the military services: Army, Navy, Air Force, and Marine Corps. Each service calculates the munitions required to cover its targets. Although each employs a unique methodology, all represent *predict-then-act* analytics. In each case, the service recommends an optimal weapons mix contingent on the planning scenarios and all the assumptions they contain.

The current MRP aims to avoid predictive failure in several ways. First, the MRP considers a small number of alternative illustrative planning scenarios that aim to approximate the range of relevant future conditions that DoD's munitions mix will need to address. Since the end of the Cold War, such scenarios have become more numerous and reflect a wider range of contingencies. Second, the process aims to reduce uncertainty by adding a wealth of details to each scenario, based on specific assumptions about global and regional security conditions, local operational conditions, various participants' alignment and force postures, choices by a range of actors involving priorities, operations, risk profiles, etc. Third, the MRP adds safety factors in the form of increased weapons requirements to compensate for uncertainties not considered in the analysis.

The MRP is often criticized for overstating requirements and for purchasing inadequate quantities of some munitions. In addition, no plan is ever foolproof, but the MRP provides little information to DoD regarding the contingencies for which its weapons mix is likely to prove sufficient and the contingencies in which it is likely to fall short. These shortcomings seem inherent in the predict-then-act approach that underlies the MRP. As noted by Loeb (2005), each illustrative planning scenario embodies a large number of assumptions, not only regarding the campaigns the United States will be called on to fight, but also on details of each campaign, such as the probability of kill (Pk) of weapons against particular targets, when forces will arrive in theater, the tactics commanders will employ, and the weather they will encounter. Typically, the analysis emphasizes a best estimate for hundreds to thousands of such factors. Efforts to reduce uncertainty by increasing the detail in the analysis can help ground some estimates in more concrete data and experience. But increasing the detail can also increase the number of factors for which the future is assumed to be similar to the past, can decrease the flexibility of the analysis and thus its ability to explore over a wide range of contingencies, and can increase the opportunities for participants in the process to tune assumptions so that the analysis generates the desired policy recommendations. In addition, although safety factors can provide robustness against the failure of some assumptions, they may fail to adequately address others. Furthermore, such safety factors may not prove to be a particularly cost-effective means to achieve robustness.

In past years, RAND has applied uncertainty-sensitive approaches to aid in defense planning.[5] These approaches include precursors to RDM that address the munitions mix chal-

---

4   Chapter 2 of Loeb (2005) provides a useful overview of the MRP.

5   See, for example, Davis and Khalizad (1996) and Camm et al. (2009).

lenge. Brooks, Bennett, and Bankes (1999) apply "exploratory analysis" to munitions planning. They characterize DoD's predict-then-act approach as a "traditional analysis" that accounts for uncertainty by performing sensitivity analyses of parameters used in the predicted scenarios. They show that the optimal solution can be very sensitive to the choice of assumptions in ways that can be confusing and misleading to planners attempting to choose portfolios of weapons. Thus, rather than search for optimal solutions, they employ the idea of satisficing solutions (Simon, 1956)—ones that are nearly but not quite optimal. Their exploratory analysis conducts a search for such satisficing solutions (using the number of days to complete the campaign as the measure of merit) in a munitions problem with three types of air-launched weapons. They find a large set of such solutions, which is invariant to the uncertainties. This set of satisficing solutions is thus significantly more robust than the optimal solution originally generated by the analysis. In addition, they argue that this set of solutions is easier to explain than any particular optimal solution, so that the set provides much more useful input into the decision making process.

Loeb (2005) applies a capabilities-based approach to PGM planning as an alternative to the prediction-based, "point scenario planning" approach currently used in the MRP. The capabilities approach that he uses finds strategies that "work across a wide variety of scenarios and uncertainties while operating within an economic framework" (p. 3). Loeb conducts an illustrative analysis of a capabilities approach to munitions planning that has four types of targets and five types of PGMs. He initially accounts for uncertainty by examining four scenarios with different mixes of targets but concludes that planning against the scenario with the most targets provides the most robust strategies. Next, he constructs five variants of this most stressing scenario, each with a different relative mix of targets. He finds that the optimal munitions portfolio for any particular scenario will fail to successfully cover all targets in the other scenarios. To identify portfolios that are more robust, Loeb randomly generates 10,000 weapons mixes for a given budget constraint and identifies the portfolios with the lowest total regret in terms of numbers of weapons used. He also explores extensions of this analysis that look at reductions in budgets and production constraints.

Davis, Shaver, and Beck (2008) looked at a similar problem of investing in capabilities to conduct global strike campaigns, which are campaigns that involve one to hundreds of weapons or, in some cases, special operations forces. They identified a number of "building blocks" to conduct global strike, which are binary choices of qualitative options, e.g., whether B-2s will be based forward and whether a conventional intercontinental ballistic missile would be developed. They created a "building blocks to composite options tool" that measured the effectiveness and cost of each possible combination of building blocks to identify an efficient frontier of options where a level of effectiveness can be reached at a minimum cost. Recognizing the uncertainty inherent in the future and the fact that many combinations of building blocks were about as effective at similar cost, they constructed "spanning sets of scenarios"[6] that they believed characterized future uncertainty. Using "portfolio analysis tools," they examined where each portfolio of building blocks was vulnerable and showed how effectiveness was calculated for each portfolio and scenario.

---

[6] A spanning set is defined as "a set of test scenarios chosen so that if alternative proposed systems are tested against them, the systems will be 'stressed' in all the appropriate ways. Systems that do well across these test cases should do well in the situations that arise in the real world" (Davis, Shaver, and Beck, 2008, p. 26).

These previous studies provide valuable insights and precedent but do not reach a level of realism or capability suitable for actual DoD planning processes. However, RDM tools and methods have advanced significantly since the Brooks, Bennett, and Bankes and Loeb studies and now may be closer to the capabilities required for practical implementation. In particular, the RDM process now emphasizes an iterative process of characterizing the vulnerabilities of proposed strategies and using this information to suggest more robust responses and the tradeoffs among them. In addition, RDM's ability to address multiple measures of merit has also improved, along with its ability to examine adaptive strategies, which achieve robustness by being designed to evolve over time in response to new information. Finally, the increased speed and memory of available computers and the power of data analysis and visualization tools has enabled RDM analyses to employ more complex models.

The present study thus expands on these previous studies by employing more detailed models, considering adaptive munitions mix strategies and more explicitly identifying the comparative vulnerabilities of these strategies. For instance, this study uses a more realistic number of weapons (30) and representative targets (30) than the previous studies. Although these mixes of weapons and targets are still an abstraction from reality, the greater variety better reflects the tradeoffs that exist between weapons in terms of cost, effectiveness, and applicability across substantially different targets.[7] By increasing this realism, this study is able to assess some of the benefits and costs of purchasing a wide variety of weapons versus focusing purchases on a subset of weapons. Second, this study considers how munitions mix strategies perform over a wide range of 20-year security environments, each filled with different types and sequences of campaigns. In contrast, the previous studies consider single campaigns. Considering dynamic, multiyear sequences of campaigns both provides a more realistic and stressing analysis of proposed munitions mix strategies and enables consideration of the ways in which munitions mix strategies might most effectively adjust over time to changing circumstances. As demonstrated in this study, the ways in which such strategies change over time often prove as important as the desired munitions mix itself in meeting U.S. national security goals.

Organization of This Report

In the remainder of this document, we demonstrate how RDM can be used to help decision makers identify robust strategies for air-delivered conventional munitions mixes. In the next chapter, we more fully describe RDM and then use the approach to scope the munitions mix planning challenge. In Chapter Three, we describe the RDM analysis, the robust munitions mix strategies it identifies, and the performance of these strategies over a wide range of plausible futures. The concluding chapter offers initial thoughts on how DoD might incorporate RDM into its planning processes.

Three appendixes support this study. The first describes the weapons on target (WoT) model used to simulate the future campaigns. The second offers more information on our experimental design, and the third presents the data used to conduct this work.

---

[7]   The 30 targets differ in their mobility, their hardness, their ability to be targeted in nonpermissive environments, the limitations they impose on weapons size to avoid collateral damage, their dispersion, and their time criticality. Table C.7, below, details these differences.

CHAPTER TWO

The RDM Approach to Munitions Mix Planning

Decision makers concerned with the proper munitions mix for U.S. forces face a significant and ubiquitous challenge. Effective policy choices require quantitative analysis of future risk and of the effectiveness of alternative munitions mix strategies. But given the fast-paced, transformative, and often surprising changes facing U.S. national security, the quantitative methods and tools commonly used to inform decisions could prove counterproductive and misleading. Any assessment of the effectiveness of future munitions mix strategies will depend on assumptions regarding the campaigns the United States is called on to wage, the capabilities of adversaries, the progress of technology, and the strategies and tactics used. But many of these factors are intrinsically hard to predict. Munitions mixes designed for one set of assumptions may prove inadequate if another future comes to pass.

Comparison of RDM and Traditional Analysis

RDM is an iterative, quantitative, decision support methodology designed to address such challenges. The approach has been applied to areas outside national security, such as flood risk (Fischbach, 2010; Lempert et al., 2013a) and water management applications (Groves and Lempert, 2007; Groves et al., 2008; Means et al., 2010) in situations where decision makers face conditions of deep uncertainty. Deep uncertainty occurs when the parties to a decision do not know—or agree on—the best model for relating actions to consequences or the likelihood of future events (Lempert, Popper, and Bankes, 2003).

RDM rests on a simple concept. Rather than using models and data to describe a best-estimate future, RDM runs models over hundreds, thousands, or even millions of different sets of assumptions to describe how plans perform in many plausible futures. The approach then uses statistics and visualizations on the resulting large database of model runs to help decision makers identify those future conditions where their plans will perform well and poorly. This information can help decision makers develop plans that are more robust to a wide range of future conditions.

This simple concept contains two particularly important ideas. First, much quantitative risk and decision analysis (in particular, DoD's planning) typically uses a *predict-then-act* approach. Analysts assemble available evidence into best-estimate predictions of the future and then use models and tools to suggest the best strategy given these predictions. Such methods work well when the predictions are accurate and not controversial. Otherwise, the methods can produce gridlock and lead to solutions that fail when the future turns out differently than expected.

In contrast, RDM runs the analysis "backward," using a *vulnerability-and-response* approach. Analysts begin with one or more strategies under consideration (often a current plan) and then, using potentially the same models and tools, characterize a spectrum of future conditions, including some where a strategy fails to meet its goals (is vulnerable). This serves as a stress-test of strategies and helps decision makers identify "robust" strategies—those that perform reasonably well regardless of what the future brings—and identify the key tradeoffs among potential robust strategies. Often, the robust strategies identified by RDM are adaptive, designed to evolve over time in response to new information (Lempert, Popper, and Bankes, 2003).

Second, traditional risk and decision analysis condenses information about a range of potential futures into a single best-estimate future (sometimes expressed as a probabilistic forecast) or a small number of planning scenarios. But RDM assembles the results of many hundreds, thousands, or even millions of computer simulation model runs and uses this database to comprehensively explore and summarize the challenges and opportunities the future might bring. In particular, RDM provides a way to effectively communicate the information in these many runs by summarizing them as a small number of decision-relevant scenarios. By embracing many plausible futures, RDM can help reduce overconfidence and the deleterious effects of surprise, systematically include imprecise information in the analysis, and help decision makers and stakeholders with differing expectations about the future nonetheless reach consensus on action (Lempert and Popper, 2005; Groves and Lempert, 2007; Hallegatte et al., 2012).

RDM Enables Decision Makers to Discover Robust Strategies Through Iteration

To implement the above concepts, RDM uses sophisticated analytic tools embedded in an explicit process of participatory stakeholder engagement (Lempert et al., 2006; Lempert and Collins, 2007). As shown in Figure 2.1, RDM follows an interactive series of steps consistent with the "deliberation with analysis" decision support process recommended by the U.S. National Research Council (2009). Deliberation with analysis begins with the participants to a decision working together to define the policy questions and develop the scope of the analysis to be performed. Subsequent steps involve expert data collection, modeling, and analysis, along with deliberations based on this information in which choices and objectives are revisited.

The RDM process begins at the top of Figure 2.1 (see Lempert et al., 2013b) with a participatory scoping activity in which decision makers define the objectives and metrics of the decision problem, strategies or options that could be used to meet these objectives, the uncertainties that could affect the success of these strategies, and the relationships that govern how strategies would perform with respect to the metrics (Step 1). This scoping activity often uses a framework called "XLRM," described below, which helps to collect and organize the information needed for the simulation modeling.

In Step 2, analysts use the resulting simulation model to evaluate the strategy or strategies in each of many plausible futures. This generates a large database of simulation model results. In Step 3, analysts and decision makers use visualizations and "scenario discovery" (Bryant and Lempert, 2010) to explore the data and identify the key combinations of future conditions in each candidate strategy that might not meet decision makers' objectives. For example, a munitions mix strategy may fail to meet U.S. goals if the security environment is more severe

**Figure 2.1
The RDM Approach**

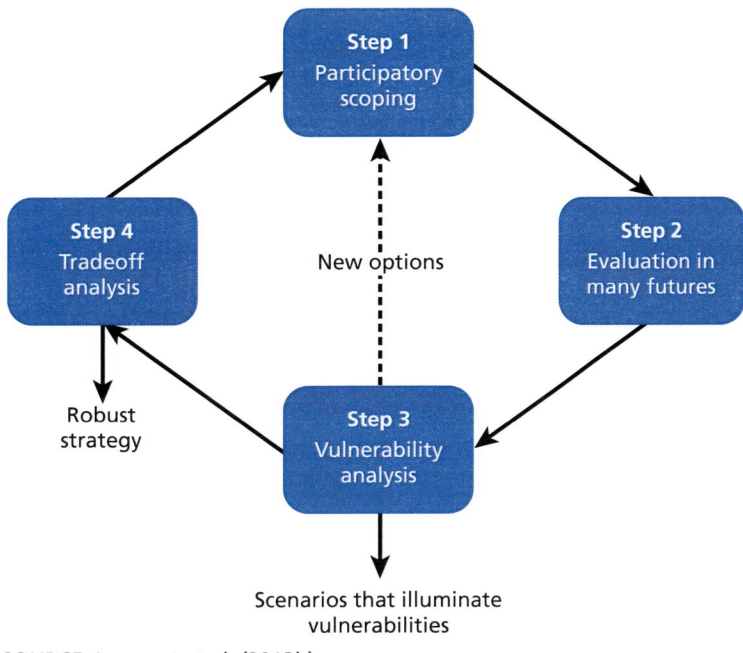

SOURCE: Lempert et al. (2013b).
RAND *RR1112-2.1*

than expected and if some key technology has less-than-hoped-for capabilities. Such a scenario ("severe security environment and poor technology performance") may concisely capture the vulnerabilities of the munitions mix strategy.[1]

Having identified a scenario in which a strategy fails to meet its goals, decision makers can turn to available evidence to consider whether the scenario is sufficiently likely as to warrant modifications to the strategy. For example, decision makers may conclude that the risk of a severe security environment and technology shortcomings is sufficiently high to justify strategy or policy changes.

These scenarios also provide the foundation for developing, evaluating, and comparing potential modifications to the alternative strategies that might reduce these vulnerabilities (Step 4). Knowing that a particular munitions mix strategy may fail to meet goals under a particular set of conditions might help decision makers decide to modify that policy with an alternative mix of weapons. Scenario discovery on this new alternative would reveal the conditions to which it is vulnerable.

Using a tradeoff analysis, decision makers may decide on a robust strategy. Or, they may decide that none of the alternative strategies under consideration proves sufficiently robust and return to the scoping exercise, this time with deeper insight into the strengths and weaknesses of the strategies initially considered.[2]

---

[1]   Specialized software tools are available to help analysts implement these steps. For example, this study used a scenario discovery toolkit to conduct the scenario discovery analysis (Bryant, 2014).

[2]   There are also other paths through the RDM process. Information in the database of model results might help identify the initial candidate strategy, or information about the vulnerabilities of the candidate strategy may lead directly to another scoping exercise to revisit objectives, uncertainties, or strategies.

As will be described in Chapter Three, this project conducted two iterations of the process shown in Figure 2.1.

XLRM Factors Shape the Design of the Experiment

As with many RDM exercises, this project employed a framework called "XLRM" (Lempert, Popper, and Bankes, 2003) to help guide model development and data gathering. Throughout the course of this study, we conducted 12 meetings with CAPE staff to review results and plan next steps in the analysis, generally using the XLRM framework to organize and document the discussions. We expect that any future RDM exercises with CAPE would also use this approach.

The letters X, L, R, and M refer to four categories of factors important to an RDM analysis:

- **Exogenous uncertainties (X)** are factors outside the control of decision makers but that may affect the ability of near-term actions to achieve decision makers' goals.
- **Policy levers (L)** are near-term actions that decision makers want to consider—in this case, the munitions they intend to purchase and how they make those purchases over time.
- **Relationships (R)**, generally represented by simulation models, describe how the policy levers perform, as measured by the metrics, under the various uncertainties.
- **Metrics (M)** are the performance standards used to evaluate whether or not a choice of policy levers achieves decision makers' goals.

In essence, RDM compares the performance of alternative combinations of policy levers, as evaluated by the metrics, over a wide range of uncertain futures using the relationships or models. Among its benefits, the XLRM framework helps distinguish among the basic elements of any good decision process: decision makers' goals, the actions they can take to achieve those goals, and external factors that might influence their ability to achieve their goals.

This chapter is organized around this XLRM framework, as summarized in Table 2.1.[3] It first describes the exogenous uncertain factors (X) that might affect the success of munitions mix strategies. It then describes the simulation models, the relationships (R) used in this project, the alternative munitions mix strategies (L), and the measures (M) used to evaluate them. A more detailed explanation of the experimental design and the models used in the study is provided in Appendix B.

Uncertainties (X)

Many uncertain factors may affect the success of a munitions mix strategy. In particular, this study aims to examine potential interplays between large- and small-scale factors. Traditional defense planning scenarios are generally differentiated by large-scale factors, such as the size of the campaign (small or large) and where it is fought (e.g., Asia, the Middle East). But each scenario also contains numerous small-scale assumptions, such as the effectiveness of particular

---

[3] Many RDM studies use the XLRM framework, although in many cases, including this study, the exposition of the factors flows more smoothly when presented in an order that is different from the framework's proper name.

Table 2.1
Factors Considered in the RDM Munitions Mix Planning Analysis

Uncertainties (X)	Policy Levers (L)
Security environment	Munitions acquisition policy
Defense funding level	Campaign guidance for each conflict
Technology impact	
Character of conflict	

Relationships (R)	Measures (M)
Analytica campaign generator (CG) of defense and operations layers	Sufficiency Success rate Days to completion of campaigns
WoT model of tactical layer	Cost 20-year acquisition

weapons against particular targets. It may not be clear the extent to which policy implications drawn from such scenarios depend on the large- or small-scale factors, or some combination.

This study thus considers many futures that explore alternative combinations of large- and small-scale factors. As described in Chapter Three, these factors will be combined together into 1,250 alternative futures. Each future will represent a specific set of assumptions about the large- and small-scale factors. The analysis will then stress-test alternative munitions mix strategies against these futures. We thus design this set of futures and the factors that constitute them to provide the most policy-relevant stress-test possible within the computational and modeling constraints of this analysis.

To characterize potential future security conditions, we first assembled a list of the 35 conflicts fought by the United States over the last century and grouped these conflicts into six categories of increasing intensity, as shown in Figure 2.2. We then constructed 25 alternative future security environments, each a 20-year sequence of different combinations of conflicts of these six security conditions, as shown in Figure 2.3. The first of these security environments represents a repeat of the last 20 years. The other 24 were chosen to provide a maximally diverse set. As described in Appendix B, we randomly generated many thousands of alternative 25-member sets of different future security environments. We then chose the set that was most diverse according to three patterns: average severities (i.e., average overall conflict intensity across all the time periods), variability (i.e., the extent of rapid switching between benign and extreme environments), and trend (i.e., the change in average intensity over time). These differing patterns seemed important to the success of munitions mix strategies and thus the most important to include in a set of security environments designed to stress-test alternative munitions mix strategies.

Each two-year period of conflict within the security environments consists of one to three military campaigns of varying intensity. We characterize each of these campaigns by varying numbers of 30 representative targets types, described in detail in Appendix C. In addition, we characterize each campaign with five attributes, representing uncertain, small-scale factors potentially important to munitions mix planning. Shown in Table 2.2, these factors were

Figure 2.2
Historical Security Conditions

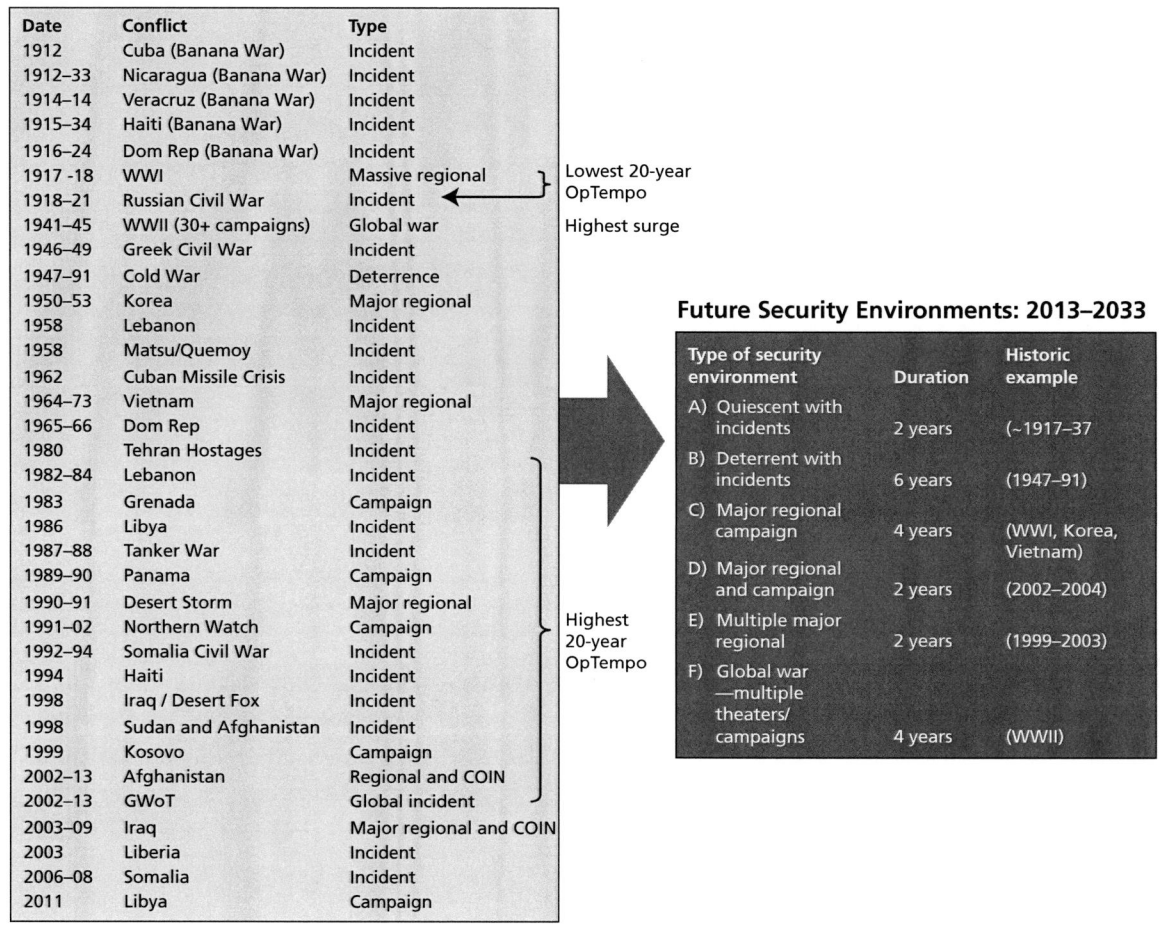

Previous Conflicts: 1912–2012

Date	Conflict	Type
1912	Cuba (Banana War)	Incident
1912–33	Nicaragua (Banana War)	Incident
1914–14	Veracruz (Banana War)	Incident
1915–34	Haiti (Banana War)	Incident
1916–24	Dom Rep (Banana War)	Incident
1917 -18	WWI	Massive regional
1918–21	Russian Civil War	Incident
1941–45	WWII (30+ campaigns)	Global war
1946–49	Greek Civil War	Incident
1947–91	Cold War	Deterrence
1950–53	Korea	Major regional
1958	Lebanon	Incident
1958	Matsu/Quemoy	Incident
1962	Cuban Missile Crisis	Incident
1964–73	Vietnam	Major regional
1965–66	Dom Rep	Incident
1980	Tehran Hostages	Incident
1982–84	Lebanon	Incident
1983	Grenada	Campaign
1986	Libya	Incident
1987–88	Tanker War	Incident
1989–90	Panama	Campaign
1990–91	Desert Storm	Major regional
1991–02	Northern Watch	Campaign
1992–94	Somalia Civil War	Incident
1994	Haiti	Incident
1998	Iraq / Desert Fox	Incident
1998	Sudan and Afghanistan	Incident
1999	Kosovo	Campaign
2002–13	Afghanistan	Regional and COIN
2002–13	GWoT	Global incident
2003–09	Iraq	Major regional and COIN
2003	Liberia	Incident
2006–08	Somalia	Incident
2011	Libya	Campaign

Lowest 20-year OpTempo

Highest surge

Highest 20-year OpTempo

Future Security Environments: 2013–2033

Type of security environment	Duration	Historic example
A) Quiescent with incidents	2 years	(~1917–37
B) Deterrent with incidents	6 years	(1947–91)
C) Major regional campaign	4 years	(WWI, Korea, Vietnam)
D) Major regional and campaign	2 years	(2002–2004)
E) Multiple major regional	2 years	(1999–2003)
F) Global war —multiple theaters/ campaigns	4 years	(WWII)

RAND *RR1112-2.2*

chosen in conversations among CAPE staff and our analysis team. Discussed in greater detail in Appendixes B and C, they include the

- effectiveness of Global Positioning System (GPS) weapons, which could become more effective or degraded by future technological changes
- ability of adversaries to blind PGMs with seeker heads
- permissiveness of adversaries' air defense, as reflected in the number of each adversary's surface-to-air missiles (SAMs) that must be destroyed before a full range of U.S. weapons can be used against targets
- adversaries' political will represented in the model as the percentage of targets that must be destroyed to defeat an adversary
- campaign operations tempo (OpTempo), represented by the delivery rates of weapons.[4]

---

[4]   To simplify the model, we assumed four broad types of delivery platforms: nonstealth aircraft, stealth aircraft, naval ships, and ground-based launchers, and we assumed that each weapon can be delivered by only one platform.

Figure 2.3
Future Security Environments Used in the RDM Analysis

	SE	2014	2016	2018	2020	2022	2024	2026	2028	2030	2032
Last 20 years	21	A	A	B	D	E	D	D	D	C	B
Benign	1	A	A	B	B	B	A	A	A	A	A
Moderate to benign	23	C	C	C	C	A	A	A	A	A	A
Benign to challenging	24	A	A	A	A	A	B	B	B	D	D
Benign+	2	A	B	B	B	E	A	A	C	C	A
Challenging to benign	22	E	E	C	C	A	A	A	A	A	A
Moderate	20	E	B	B	B	A	A	B	B	B	E
Volatile	25	A	E	A	E	E	A	E	A	A	A
Moderate (steady)	4	B	B	B	D	D	B	B	B	D	D
Moderate (spikes)	3	A	E	B	B	B	E	A	C	C	E
Moderate	17	C	C	F	F	A	B	B	B	B	B
Benign to challenging	19	A	B	B	B	B	B	B	F	F	D
Difficult (steady)	5	E	D	D	C	C	B	B	B	C	C
Difficult (surge)	6	C	C	F	F	B	B	B	A	A	E
Difficult (surge) 2	8	C	C	B	B	B	F	F	A	A	E
Difficult	14	E	B	B	B	D	D	E	A	C	C
Difficult (increasing)	13	B	B	B	B	B	B	F	F	D	D
Moderate	16	C	C	B	B	B	E	C	C	E	D
Difficult (varied)	10	D	D	A	E	A	F	F	B	B	B
Difficult (varied, increasing)	12	A	A	A	F	F	E	A	D	D	D
Challenging +	7	F	F	C	C	B	B	B	C	C	E
Challenging + 2	9	C	C	F	F	B	B	B	C	C	E
Challenging + 3	11	D	D	A	A	F	G	E	C	C	B
Challenging (increasing)	15	B	B	B	F	F	D	D	A	D	D
Challenging (increasing)	18	E	B	B	B	A	D	D	F	F	D

NOTES: Security environments (SEs) are ordered so that the last 20 years is listed first, and then entries are ordered according to increasing severity. SE A is the most benign and SE F is the most stressful.

RAND *RR1112-2.3*

Table 2.2
Uncertain Campaign Attributes

Weapons characteristics	GPS technology	Performance of GPS weapons improves or degrades by ±50%
Weapons characteristics	Blinding (B)	Performance of PGMs with seeker heads degrades by 0% to 75%
Adversary capability	Permissiveness of air defense (PERM)	Number of SAM targets increases or decreases by ±50%
Adversary political will	Adversary's political will (WILL)	Adjust the percentage of targets destroyed required to end campaign (normally 80%) by ±20%
OpTempo	Delivery rate (DR)	Adjust delivery rates (both surge and steady-state) by ±50%

Relationships (R): Models

We built two coupled models for this study to link policy choices to outcomes, as shown in Figure 2.4. The WoT model uses weapons inventories to fight individual campaigns. The CG generates 20-year sequences of campaigns and provides the munitions to fight them. The two models are configured so that we can run thousands of alternative futures for each of several munitions mix strategies.

The WoT model simulates individual campaigns on a day-to-day basis, using an optimization algorithm to match munitions and delivery vehicles to targets.[5] WoT uses stochastic Pks to determine whether targets have been destroyed. WoT is similar to such campaign models as the Air Force's Combat Forces Assessment Model and related models used by CAPE, but with less detail. WoT's simplicity provides significant advantages in run-time. It can run a large, complex campaign in at most a few minutes on a modern desktop computer, compared to hours or days for a more complex campaign model. However, WoT cannot simulate campaigns with the same fidelity as that of more detailed models.

WoT requires inputs that specify the details of each campaign. For example, it requires a target set with different attributes for the targets, a weapons portfolio with different attributes for each weapon, and a Pk table that describes the likelihood that any weapon will destroy any given target. Appendix A provides a detailed overview of how WoT works and its data requirements. WoT uses probabilistic Pks, so the analysis uses 100 stochastic runs of WoT for each campaign and reports summary statistics on the outcomes.

The CG provides a series of campaigns and their attributes to the WoT model. Each case considered in the RDM analysis consists of one munitions strategy tested against one future. The CG constructs a future by beginning with one of the security environments in Figure 2.3 and a specific set of values for the campaign attributes in Table 2.2. Together, these constitute the uncertainties (Xs) described above. As described in the next section, we use a statistically chosen experimental design to create the thousands of combinations of alternative security environments and campaign attributes. For each future and using one of the munitions strategies—the policy levers (Ls) described below—the CG sends one campaign at a time to WoT, along with information on the available munitions. WoT reports back the campaign outcomes, including the number of weapons used. The CG adjusts U.S. weapons inventories, purchases additional weapons using available funding, and the sends WoT another campaign.

Figure 2.4
Flow of Information Between the WoT Model and the CG

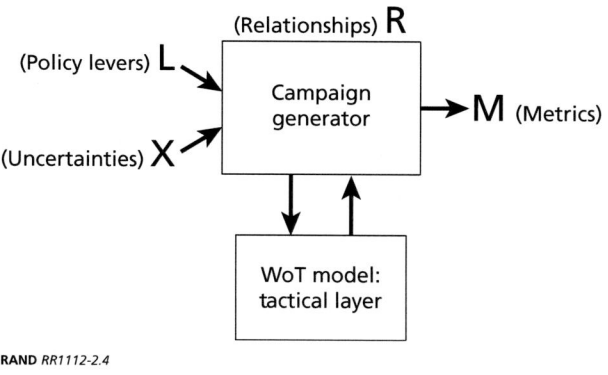

RAND RR1112-2.4

---

[5] WoT's treatment of delivery vehicles and munitions is currently highly simplified. WoT assumes that each munition can be delivered by only one type of four broad classes of delivery vehicle (nonstealth and stealth aircraft, sea, or land). In addition, WoT does not reserve any vehicles or munitions for day-to-day noncontingency use. Increases in fidelity of how WoT simulates delivery vehicles will be considered for future research, especially when real data are used.

At the end of the 20-year sequence the CG summarizes the outcomes, using the metrics (Ms) discussed below.

Policy Levers and Strategies (L)

In practice, most policies are adaptive, adjusting over time in response to new information. Much less often are policies designed to be adaptive. A strategy so designed takes near-term actions with explicit consideration of how they might be subsequently adjusted. A strategy designed to be adaptive may also include systematic consideration of how it will gather information and respond to it in the future.

Traditionally, analyses of munitions mix strategies have not considered how such mixes might adjust over time. Rather, they focus on describing some single best mix at a single point in time. This study takes a first step toward considering adaptive munitions mix planning strategies. We characterize each strategy with two components:

- **Desired portfolio goals** that specify the number of each type of munitions that policy-makers would like to have in the U.S. stockpile. These portfolio goals are related to the alternative force sizing constructs often used in DoD planning.
- **Purchasing rules** that describe how munitions will be purchased to replenish the stockpile when it is depleted during one or more campaigns.

This study considers a variety of alternative desired portfolio goals. We consider a baseline portfolio, which aims to represent the current DoD portfolio.[6] In addition, we consider a variety of alternative portfolios, constructed by running an optimization routine with the WoT model for specified campaigns and objective functions.

During the course of a simulation run weapons will be expended across the various campaigns. Our munitions mix strategies thus also include purchasing rules for replenishing the stockpile. We assume that total defense funding varies with the severity of the security environment in the previous period, representing the historically observed lag times between changes in observed threat and changes in the defense budget. We assume that 0.85 percent of defense funding is available for munitions purchases, consistent with historical averages. During each time period, the strategy allocates the funding available to purchase munitions to replenish the stockpile based on alternative purchasing rules. For instance, the strategy might prioritize weapon purchases on the basis of the number of each weapons type used in the last campaign or on the ratio in which those weapons appear in the desired munitions mix.

As a result of these rules, the munitions mix at any particular time in any particular simulation run depends on how the future unfolds. As suggested by Figure 2.5, in some futures, the munitions mix stays close to the desired numbers of each type of munition. In other futures, it may deviate significantly. As discussed in Chapter Three, the rules for purchasing weapons after they are depleted can be as important to the success of a munitions mix strategy as the desired portfolio goals themselves.

---

[6]   All weapons data in this study were taken from unclassified sources or were estimated after reviewing information in the unclassified literature. Greater fidelity of weapons attributes could likely be achieved by using classified data sources that are available to DoD. Furthermore, by limiting the study to 30 weapons and 30 types of targets, the study may have overlooked other combinations of weapons and targets that DoD planners recognize as important but that are not well known in open sources.

Figure 2.5
Evolution of an Adaptive Strategy

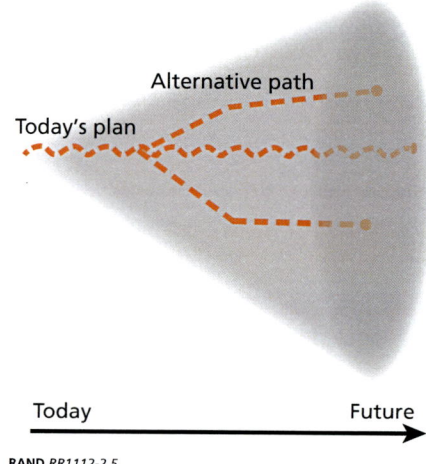

RAND *RR1112-2.5*

This study did not consider in its relationships or in its policy levers a wide range of munitions industrial base issues that might prove relevant, in particular to the adaptive strategies considered here. For instance, we did not consider policies that might make it possible to more quickly increase production of some munitions or how the size or rate of purchase might affect costs. Our analysis does impose a two-year latency on all munitions purchase decisions, which may soften the effect of these assumptions. However, expanding this analysis to include such industrial base issues could be a valuable direction for future work.

Measures (M)

To compare the performance of alternative munitions purchasing strategies in each of the many futures, this RDM analysis uses two measures: cost and military sufficiency.

Cost is measured in two ways. For each simulation run, cost is measured as the purchase cost of all the weapons expended across all campaigns in a simulation run. This includes the purchase cost of the existing munitions expended throughout the run and the purchase cost of newly purchased munitions expended throughout the run. More generally, the cost of each munitions purchasing strategy can be compared through the acquisition cost of each strategy's desired munitions portfolio, which is the total cost of purchasing all weapons in the portfolio.

The analysis uses two measures of military sufficiency: success rate and days to completion. The success rate records the percentage of campaigns won. Campaigns are won when the required targets are destroyed within a maximum campaign length of 200 days.[7] Days to completion records the number of days required to win a campaign. A shorter time is desirable because it reduces other, nonmunitions-related costs of the campaign.

In general, decision makers would prefer a low-cost munitions mix strategy that can win all campaigns and win them quickly, across a wide range of plausible futures. However, as discussed in the next chapter, there are often tradeoffs among these objectives.

---

[7]   A limit of 200 days was placed on each campaign. If a campaign went over 200 days, it was deemed unsuccessful. In practice, this 200-day limit was rarely binding; campaigns that lasted 200 days had usually run out of usable weapons.

It is worth noting that although success rate and days to campaign completion correspond to key goals of Pentagon leaders, analysts and decision makers often distrust these two measures as useful ways to quantify military sufficiency. In part, this distrust owes to the obvious difficulty of predicting these quantities with any confidence. However, using success rate and days to campaign completion in an RDM analysis may render them more useful because such an analysis does not predict, but rather evaluates, how these quantities depend on a wide range of assumptions and alternative policy choices.

CHAPTER THREE

RDM Munitions Mix Analysis

We can now use the simulation model and data presented in Chapter Two to conduct the munitions mix RDM analysis. We conducted two iterations of the loop shown in Figure 2.1 during the course of the project. The first iteration focused on a broad array of strategies developed to address relatively simple planning scenarios. The second iteration, whose strategies were based on findings from the first, focused on a narrower range of strategies more specifically designed to be robust over a wider range of futures. This chapter will briefly summarize the analysis of the first set of strategies and then will present the analysis of the second set in more detail.

Initial Analysis of a Broad Range of Munitions Mix Strategies

As described in the previous chapter, we consider adaptive munitions mix strategies that have both a *desired portfolio goal* that specifies desired number of weapons of each type and a *purchasing rule* that specifies the order in which munitions will be purchased when the stockpile is depleted. In the initial scoping of strategies (Step 1), we considered five portfolio goals and two purchasing rules for a total of ten strategies. We choose these goals and strategies, in consultation with CAPE, to provide a broad scan of the performance of alternative strategies.

The five portfolio goals are derived from alternative DoD force sizing constructs. In particular, we considered

- *Baseline*, representing the current munitions portfolio
- *Small-Cost*, which purchases weapons designed to fight—at least cost—a Small War planning scenario with prolonged irregular campaigns
- *Small-Time*, which purchases weapons designed to fight—as quickly as possible—a Small War planning scenario with prolonged irregular campaigns
- *Big-Cost*, which purchases weapons designed to fight—at least cost—a Big War planning scenario with two back-to-back, medium-size campaigns
- *Big-Time*, which purchases weapons designed to fight—as quickly as possible—a Big War planning scenario with two back-to-back, medium-size campaigns.

For all but the *Baseline*, the study used the WoT model to find the optimum munitions portfolio for the specified planning scenario (Small War or Big War) and objective function (least cost or least time). This process—with its planning scenarios and optimization goals—

aims to approximate the current DoD predict-then-act approach to munitions planning. The appendixes of this report provide more details on these calculations.

For the two purchasing rules, we considered

- *Stockpile*, which spends all available funding in proportion to the portfolio goals, and continues purchasing weapons as long as funding is available, and
- *Replenishment*, which purchases only until the desired portfolio goal has been reached. If funding is insufficient to fully meet the portfolio goals, the replenishment rule spends in proportion to the shortages in the goals.

Note that the stockpile rule can continue purchasing weapons beyond the portfolio goals, so in some futures, many more weapons may be purchased than the replenishment goal. The replenishment rule focuses munitions purchases on replacing weapons that were heavily used in the previous period; that is, the rule shifts purchasing priorities dynamically as information is learned about which weapons have proven most useful to combat commanders. Even with the same portfolio goal, the two purchasing rules can lead to very different weapons inventories.

The team next evaluated the ten strategies over a wide range of plausible futures (Step 2). Each future consisted of one security environment and one combination of each of the small-scale uncertainties shown in Table 2.2. The computational resources available to this study gave us the ability to consider analyses with about 10,000 total cases, representing about a week of run time on RAND's ten-core cluster computer. We allocated those 10,000 cases by pairing 20 security environments with 50 alternative combinations of assumptions about the small-scale uncertainties,[1] so that each of the ten strategies was tested in 1,000 different futures.

This database of runs supported an initial vulnerability analysis of the ten strategies, which RAND and CAPE staff used to refine the strategies considered in the second iteration of the analysis. In particular, this initial vulnerability analysis suggested the following.

First, the strategies based on the Small War planning scenarios, *Small-Cost* and *Small-Time*, had very poor success rates (that is, they rarely destroy the required number of targets) in any but the least-stressing security environments. Not surprisingly, these strategies seemed appropriate only if defense planners were confident that larger contingencies were exceedingly unlikely.

Second, the strategies with stockpile purchasing rules generally outperformed the strategies with replenishment purchasing rules, as might be expected because the stockpile rule purchases more munitions. But, surprisingly, in many futures, the replenishment rule outperformed the stockpile rule. The replenishment rule performed better in those futures in which a series of campaigns closely spaced in time would significantly deplete weapons inventories and insufficient funding was available to replace them quickly. In such futures, the replenishment rule spent its limited funding in a smarter way. It focused its funding on munitions that had been heavily used in the previous period, whereas the stockpile strategy bought a constant proportion of weapons no matter what had been used in the previous period.

Third, the cost-minimizing portfolio goal, *Big-Cost*, often had higher success rates than the time-minimizing portfolios' goals, *Big-Time*. We expected that the *Big-Cost* strategy, which purchases fewer weapons that are less capable than does *Big-Time*, would trade lower cost for

---

[1]   The 50 combinations were selected using a Latin Hypercube experimental design that randomly samples parameter values but spaces them to ensure a relatively uniform distribution of parameters.

less success and longer campaign completion times. This did occur in some futures. In some futures, both *Big-Cost* and *Big-Time* had high success rates. In these futures, *Big-Time* completed the campaigns more quickly. In some futures, *Big-Cost* failed to complete campaigns because it ran out of standoff weapons before it rendered the security environment permissive. But surprisingly, in a relatively large share of futures, *Big-Cost* completed campaigns that *Big-Time* did not. In such futures, repeated campaigns would deplete weapons inventories. *Big-Cost* strategies, with their less-expensive weapons, had sufficient funding to replenish their stockpiles before the next campaign began. *Big-Time*, with its more expensive weapons, could not and was thus unready to complete the campaign.

Fourth, the vulnerability analysis identified two types of futures in which both *Big-Cost* and *Big-Time* performed poorly. These strategies proved vulnerable in futures with degraded GPS performance. These strategies also failed against deterrence campaigns, which, although relatively small, require high-performance weapons. Deterrence campaigns (characterized by numerous short-duration campaigns, employing high-tech standoff PGMs at a high usage rate) are thus not lesser-included cases of the Big War planning scenarios.

Analysis of Potentially More Robust Munitions Mix Strategies

We next developed a new set of alternative strategies in consultation with CAPE and using the findings from the first round of vulnerability analysis. As in the initial analysis, each strategy consists of desired portfolio goals and purchasing rules. However, the portfolio goals for these strategies were specifically designed to be robust over a wider range of futures.

In particular, we used the WoT model to find portfolios with optimal performance in each of several carefully constructed planning scenarios, each designed to address specific vulnerabilities identified in the initial analysis. We also eliminated the Small War portfolios, which performed too poorly in the initial analysis to warrant continued consideration.

Both the *Big-Time* and *Big-Cost* portfolios from the initial analysis had poor performance in deterrence campaigns, which, although smaller in size, did not represent a lesser-included case of the Big War planning scenario. Each of the three additional strategies thus added a deterrence campaign to the Big War planning scenario. In addition, rather than optimizing portfolio goals for either minimum cost or time, we optimized for various mixes of these two objectives. As described in Appendix B, the *Big+Deter-Time* strategy, heavily weighted (95 percent) toward the time-minimized portfolio, provided the best mix of the existing time- and cost-minimized objectives. The two other new strategies address specific vulnerabilities identified in the initial analysis. *Big+Deter-Mixed* addresses the vulnerability of cost-minimized portfolios to running out of expensive standoff weapons by using planning scenarios where it can employ only standoff weapons. This increases the costs but pushes the strategy to use cost-effective standoff weapons. *Big+Deter/GPS-Mixed* addresses the vulnerability to low GPS effectiveness by using a planning scenario with a 33 percent reduction in the effectiveness of GPS weapons, which results in a strategy with larger inventories of munitions.

For this new set of alternative strategies, we also used only the replenishment purchasing rule from the initial analysis, in which weapons are purchased in proportion to *shortages* in portfolio goals. This new set of strategies did not include the stockpile purchasing rule, in which weapons are purchased in proportion to portfolio goals. The initial analysis suggested the replenishment purchasing rule performed better than the stockpile purchasing rule in many futures because the former purchased the weapons most heavily used in the previous conflicts. In addition, the one advantage of the stockpile rule—that it continued to purchase weapons beyond the portfolio goals as long as funding was available—seemed unrealistic to defense planners at CAPE.

Table 3.1 summarizes the six alternative munitions mix acquisitions strategies considered in this stage of the analysis. The first three—the *Baseline* portfolio and the two Big War portfolios that optimized for least cost and least time to finish the campaign—remain from the initial analysis. The last three are the new "mixed" portfolios. Note that the estimated cost of all the portfolios is similar, with the exception of *Big-Cost*, which is considerably less expensive.

Figure 3.1 compares the relative size and mix of munitions in terms of estimated acquisition costs (i.e., the cost if the entire portfolio were purchased at once) for three of the portfolios in Table 3.1. The left panel shows the estimated acquisition value of current baseline inventories, the middle panel shows the *Big-Cost* portfolio, and the right panel shows the *Big+Deter-Mixed* portfolio. Note that the *Big+Deter-Mixed* and especially the *Big-Cost* portfolios have significantly less diversity of weapon types. In general, the portfolios developed in this study use fewer types of munitions than the current *Baseline* portfolio and usually concentrate most acquisition spending on a handful of munitions, for two primary reasons. First, the baseline portfolio has evolved over time as new weapons have been developed, but older weapons remain in inventories. Second, DoD's planning scenarios are likely more detailed than the 30 target types[2] in this study, thus providing more opportunities to match specialty weapons to particular target types.[3] In the final chapter of this study, we will discuss some implications of these differences in weapons mix diversity.

Table 3.1
Strategies Used in the RDM Analysis

Strategy Name	Explanation	Estimated Acquisition Cost ($ Billions)
Baseline	Desired portfolio goal is *Baseline* existing inventories	73.1
Big-Cost	Desired portfolio goal minimizes the *cost* of campaigns	14.4
Big-Time	Desired portfolio goal minimizes the *time* of campaigns	66.7
Big+Deter-Time	*Deterrence* campaign included in planning scenarios; reduces total costs with a *95/5* ratio of time-minimizing portfolio to cost-minimizing portfolio	67.2
Big+Deter-Mixed	*Deterrence* campaign included in planning scenarios; cost-minimizing portfolio heavy in standoff weapons; reduces total cost with a 55/45 ratio of time-minimizing portfolio to cost-minimizing portfolio	60.6
Big+Deter/GPS-Mixed	*Deterrence* campaign included in planning scenarios with GPS degradation; reduces total cost with a 75/25 ratio of time-minimizing portfolio to cost-minimizing portfolio	84.5

NOTES: The Big Wars strategies include weapons to fight two back-to-back, medium-size, major regional campaigns. All strategies in this table use the replenishment purchasing rule.

---

[2] See Table C.7 for a complete listing and characterization of target types employed in the project.

[3] See Table B.1 for a complete listing of munitions employed in the project.

Figure 3.1
Relative Acquisition Costs of Weapons in Three Desired Portfolio Goals

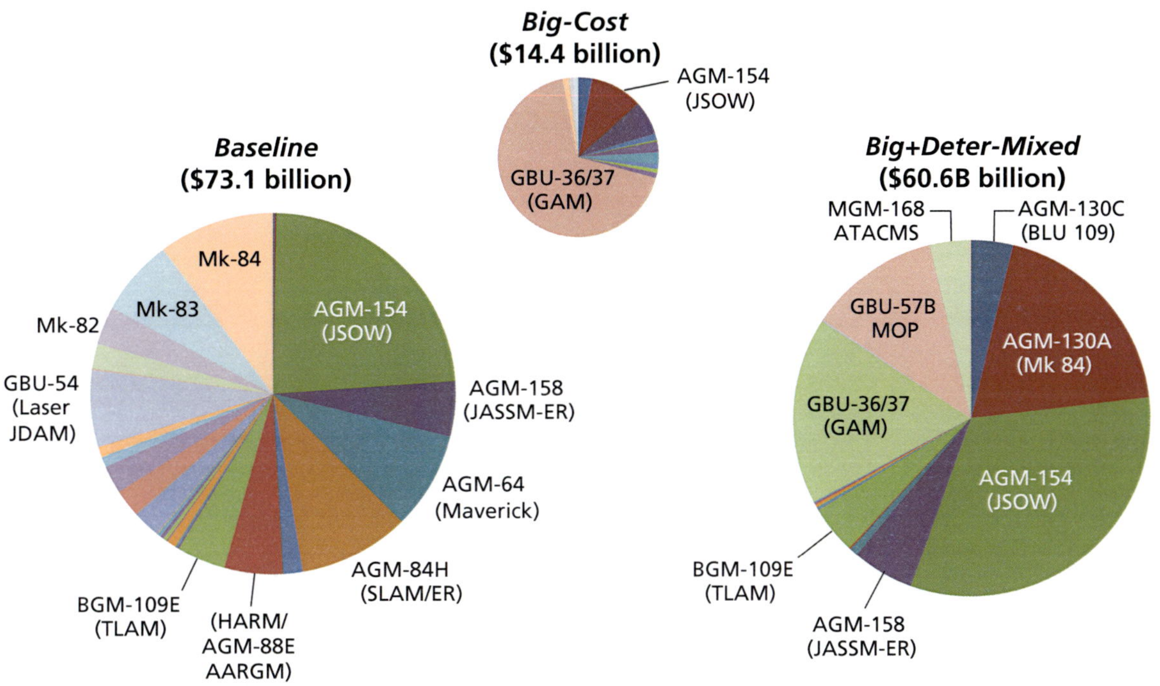

NOTE: Weapons with an estimated acquisition cost of over $2 billion are labeled.
RAND *RR1112-3.1*

Stress-Testing of Strategies over Many Futures

The team next evaluated the six strategies over a wide range of plausible futures (see the box labeled "Evaluation in many futures" in Figure 2.1). Each future consists of one of the 25 security environments (the large-scale factors shown in Figure 2.3) and one combination of values for each of the small-scale uncertainties shown in Table 2.2. We paired 25 security environments[4] with 50 alternative combinations of assumptions about the small-scale uncertainties generated with a Latin Hypercube experimental design, so that each strategy was tested in 1,250 different futures. With its six alternatives strategies, this stage of the analysis required 7,500 (6 × 1,250) total cases, which took about a week of run time under ideal conditions on RAND's ten-core cluster computer.[5]

Figure 3.2 provides an initial screening of the tradeoffs among the six alternative strategies. The figure compares the success rates and costs for each strategy averaged over the 1,250 futures.[6] The best strategies are those with relatively high success rates and relatively low

---

[4] The second iteration of the RDM analysis added five additional security environments to the 20 used in the first iteration. One of these additional security environments approximated the security environment of the past 20 years, and the other four focused on less-severe security environments that were underrepresented in the original 20 security environments.

[5] In practice, each iteration of the analysis took much longer than a week. Other applications competed for resources on the computer, which limited the number of cases that could be run at once and slowed the simulations. Furthermore, some programming errors were not apparent until the cases were run; therefore, cases sometimes ran multiple times.

[6] The RDM analysis also tracks days to completion as one measure (M). However, at this stage of the analysis, days to completion has limited value because the strategies have a large divergence in success rates. The "marginal" campaigns

Figure 3.2
Average Success Rates and Average Costs of Munitions Expended Across Strategies

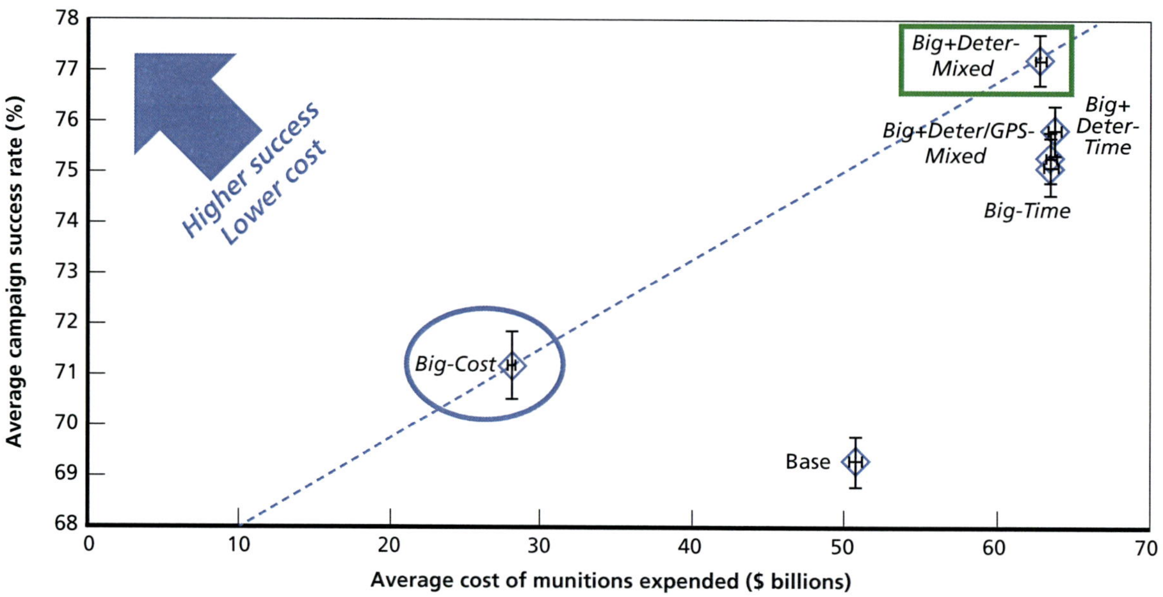

NOTES: Average campaign success rate is based on the number of successful trials throughout each 20-year simulation averaged across all 1,250 futures. Average cost is the cost of munitions expended (rather than purchased) throughout each 20-year simulation averaged across all 1,250 futures. This cost can include the cost of weapons that are already in inventories at the beginning of scenarios. Error bars are ±1 standard error. Standard errors were estimated using 100,000 bootstrap samples each of the average success rate and average cost of each case. The reader is cautioned that although some of the error bars overlap, all of the differences between strategies are statistically significant, since the strategies' performances are highly correlated.
RAND RR1112-3.2

costs. The cluster of points in the upper right of the figure suggests that the *Big+Deter-Mixed* strategy performs better than *Big-Time* and the two other mixed strategies. *Big+Deter-Mixed* has a higher average success rate than *Big-Cost*, but the latter costs half as much. The *Baseline* strategy performs less well than the others. The *Big-Cost* and *Big+Deter-Mixed* strategies thus suggest a tradeoff frontier. Decision makers might choose between these two strategies depending on how much they valued success rates relative to costs.[7]

Scenarios That Illuminate the Vulnerabilities of Strategies

After reviewing the initial screening in Figure 3.2, the team decided to examine in more detail the strengths and potential weaknesses of the *Big+Deter-Mixed* strategy. This choice reflects a judgment that senior decision makers may regard campaign success rate as more important

---

where some strategies are successful but others are not successful tend to be stressful and necessitate a large number of days to completion; thus, strategies that are successful would be penalized for having a large number of days to completion.

[7]   The reader should note that the cost differences shown in Figure 3.2 are statistically significant. The error bars show the standard errors over the stochastic simulations treated in the analysis. In general, the reported differences among the performance of strategies in this analysis have high statistical significance because each strategy is tested against the same set of futures, and the rankings of strategies are consistent across these futures, even when the percentage differences are relatively small.

than cost. In addition, in some futures considered in the analysis, one of the other strategies, for instance *Big+Deter/GPS-Mixed*, performs better than *Big+Deter-Mixed* or *Big-Cost*. But, as shown in the following discussion, these two strategies dominate the others over a sufficiently broad range of futures and performance measures that it makes sense to focus the vulnerability and response option analysis on the *Big+Deter-Mixed* strategy.

Although *Big+Deter-Mixed* has the highest average success rate of the six alternative strategies, its average rate (77 percent) is not as high as one would like. We thus ask the question: What key characteristics distinguish the futures in which *Big+Deter-Mixed* has a high success rate from those futures in which it does not? In consultation with CAPE, the team defined an acceptably high success rate as 90 percent or greater.

Applying scenario discovery algorithms (Bryant and Lempert, 2010) to the database of 1,250 futures suggests that of all the uncertainties considered in this analysis, three—two large-scale and one-small scale—are most important in distinguishing those futures where *Big+Deter-Mixed* has a high success rate. These are the following:

- **Average severity** measures the severity of the security environment, normalized so that 20 straight years of quiescent security conditions is 0, and 20 straight years of global war is 1. There is a strong relationship between average severity and success; high-severity security environments have too many targets for any strategy to be highly successful.
- **Trend** measures how average severity changes over time. A value of 0.5 means that the severity stays constant over time; higher values indicate that the severity increases, and lower values indicate that the severity decreases. The *Big+Deter-Mixed* strategy was vulnerable to futures in which the trend was far from constant.[8]
- **GPS** measures how the effectiveness of GPS weapons changes over the baseline, in terms of a percentage (i.e., 0.05 is a 5 percent improvement in effectiveness). *Big+Deter-Mixed* has less success when GPS is degraded. With improvements in GPS weapons, *Big+Deter-Mixed* can be more successful in futures that are more severe.

Figure 3.3 displays the two scenarios defined by these three uncertain parameters. In the Moderate Scenario—shown in the figure as the region inside the black, dashed line— are futures in which the *Big+Deter-Mixed* strategy has a success rate greater than 90 percent. In the Extreme Scenario, *Big+Deter-Mixed* has a success rate less than 90 percent. Of the 1,250 futures considered in our analysis, 361 fall into the Moderate Scenario, whereas the other 889 futures fall into the Extreme Scenario. Note that the security environment over the past 20 years (dark circles in Figure 3.3) lies just inside the Moderate Scenario as long as GPS effectiveness remains high.

These two scenarios—Moderate and Extreme—suggest that in security environments up to the severity level of the last 20 years, the *Big+Deter-Mixed* munitions acquisition strategy will have generally high success rates over a wide range of assumptions about other uncertainties, assuming a high level of GPS effectiveness. If GPS effectiveness is low, the *Big+Deter-Mixed* strategy will have high success rates only in security environments about half as severe as those in the past 20 years. The *Big+Deter/GPS-Mixed* strategy can mitigate this sensitivity

---

[8]   Since only two points in Figure 3.3 are excluded by the restrictions on the trend, it is possible that the trend was serving as a proxy for some other feature of those two security environments that made them particularly stressful.

Figure 3.3
Construction of the Moderate Scenario: Futures in Which *Big+Deter-Mixed* Has High (Greater Than 90 Percent) Success Rate

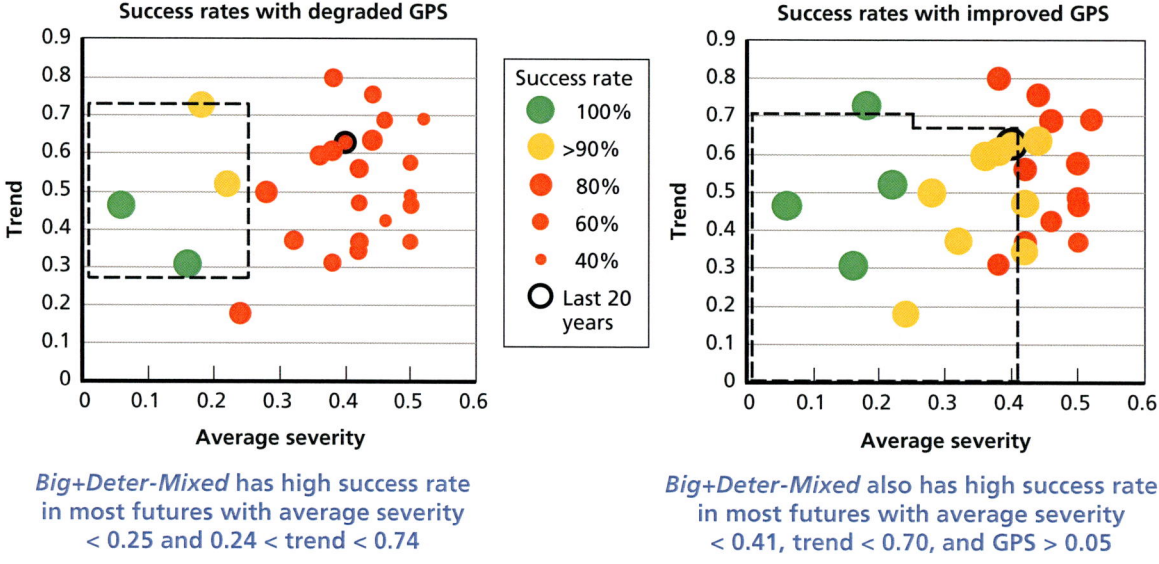

Big+Deter-Mixed has high success rate in most futures with average severity < 0.25 and 0.24 < trend < 0.74

Big+Deter-Mixed also has high success rate in most futures with average severity < 0.41, trend < 0.70, and GPS > 0.05

NOTES: Each point represents the average success rate for a security environment across a variety of futures. Each security environment is included in 50 of the futures—23 with GPS > 0.05 and 27 with GPS < 0.05.
RAND RR1112-3.3

to degraded GPS. But as Figure 3.2 shows, this strategy trades increased success when GPS is degraded for decreased success in other types of futures.

Later in this chapter, we will briefly discuss how further RDM analyses might suggest munitions acquisition strategies that improve performance in Extreme futures outside the Moderate Scenario. But now, we turn to a more focused examination of the tradeoffs among strategies in the Moderate Scenario.

Performance of *Big+Deter-Mixed* Strategy in the Moderate Scenario

Figure 3.4 compares the six alternative strategies using the maximum regret across all the futures in the Moderate Scenario[9] for the three measures of interest in this study: success, the cost of the munitions expended, and days to completion.[10] Figure 3.2 focused on average

---

[9]   For a particular purchasing strategy, regret is the difference between that strategy's measure and the measure of the best strategy. For example, if Strategy A takes 50 days to complete a campaign and Strategy B takes 60 days, the regret of Strategy A is 0 days (it is the best strategy) and the regret of Strategy B is 10 days (it takes 10 days longer than the best strategy).

This study uses a modified calculation of regret that looks only at regret to other strategies that are at least as successful at conducting a particular campaign. Thus, if Strategy A spends $80 billion and has 100 percent success, Strategy B spends $90 billion and has 100 percent success, and Strategy C spends $40 billion and has 50 percent success, the regret of Strategy A is $0 (no strategy spent less and was as successful), the regret of Strategy B is $10 billion (Strategy A was $10 billion cheaper and had the same success), and the regret of Strategy C is $0 billion (no strategy spent less and was as successful).

[10]   Days to completion regret is calculated based on only the 172 futures in the scenario where all strategies were 100 percent successful. As noted in an earlier footnote, the days to completion metric will provide a penalty for strategies that take a long number of days to complete stressful scenarios even though other strategies are unsuccessful. Calculating days to

Figure 3.4
Maximum Success Regret, Maximum Days to Completion Regret, and Maximum Adjusted Cost Regret in the Moderate Scenario

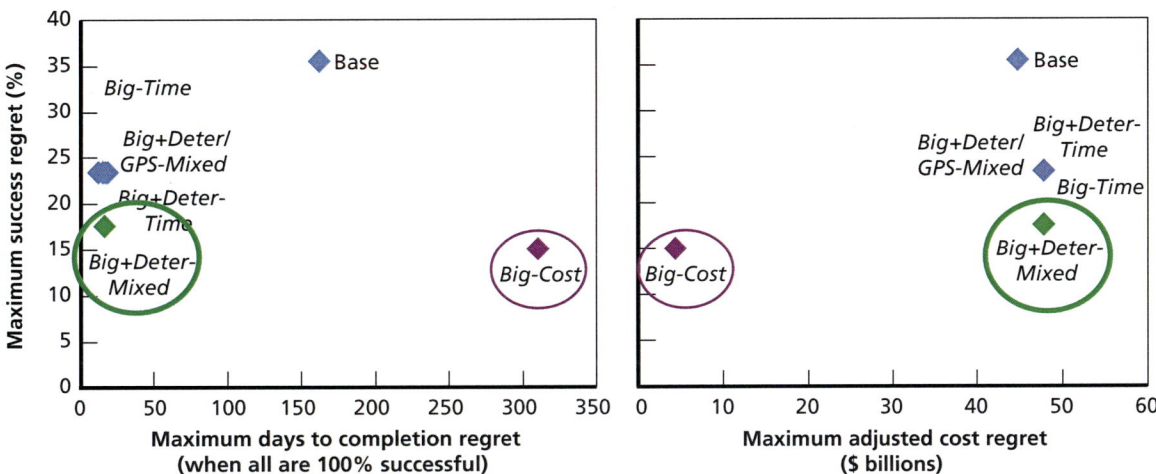

NOTE: Days to completion regret was calculated only for futures where all of the strategies were successful 100 percent of the time (172 futures).
RAND RR1112-3.4

performance, but this figure focuses on worst-case performance. Nonetheless, the basic patterns remain the same across the two figures. *Big+Deter-Mixed* generally completes the most campaigns (second-lowest success regret) and generally completes them the fastest (among the lowest days to completion regret). *Big-Cost* also has high success rates and is the least costly. However, the *Big-Cost* strategy can take much longer to complete campaigns relative to *Big+Deter-Mixed* and the other strategies, as shown by its high days to completion regret.

Table 3.2 summarizes the performance of the six alternative strategies across the 361 futures in the Moderate scenario using all the measures considered in the analysis. *Big+Deter-Mixed* and *Big-Cost* dominate for all but one measure (*Big+Deter-Time* and *Big-Time* each has slightly smaller maximum days to completion regret than does *Big+Deter-Mixed*). *Big+Deter-Mixed* has zero success regret in 323 of the 361 futures. *Big-Cost* is not far behind, with zero success regret in 313 of the futures. With its much lower cost, *Big-Cost* has both zero cost regret and zero success regret across 277 futures.

Figure 3.5 focuses on the strengths and limitations of *Big+Deter-Mixed* by plotting its days to completion and success regret across all 361 Moderate Scenario futures. Because many dots overlap, the "whiskers" indicate the number of overlapping dots. The inset table shows the distribution of futures at the origin (zero success and days to completion regret) where the number of overlapping dots is particularly high.

Not surprisingly, *Big+Deter-Mixed* has zero success regret and days to completion regret in most futures in the Moderate Scenario. In 208 of these 361 futures, *Big+Deter-Mixed* has zero regret for both measures; that is, it has a success rate and days to completion at least as good as the other five strategies. However, it is important to note that zero regret does not necessarily

---

completion only for futures where all strategies are 100 percent successful ensures that the measure provides a comparison of futures in which the strategies have the same level of success.

Table 3.2
Summary of Measures in the Moderate Scenario (361 Futures)

Strategy	Maximum Adjusted Cost Regret ($ Millions)	Maximum Success Regret (%)	Maximum Days to Completion Regret (When All Are 100% Successful)	Number with Zero Success Regret	Number with Zero Success and Cost Regret	Average Cost ($ Millions)	Average Success (%)	Average Days to Completion (When All Are 100% Successful)
Baseline	44,776	36	162	186	38	34,526	91.3	104
Big-Cost	4,315	15	310	313	277	19,745	95.5	124
Big-Time	47,848	24	15	277	12	49,300	94.8	72
Big+Deter-Time	47,842	24	13	281	9	49,310	95.0	72
Big+Deter-Mixed	47,849	18	16	323	17	48,934	95.9	72
Big+Deter/ GPS-Mixed	47,845	24	17	295	8	49,620	95.3	73

NOTES: Days to completion was calculated only for futures where all of the strategies were successful 100 percent of the time (172 futures). Shaded rows indicate the two best-performing strategies.

imply a high success rate. In 25 of the futures at the origin of Figure 3.5, *Big+Deter-Mixed* has success rates less than 90 percent, although it still has success rates at least as high as the alternative strategies.[11] In a small number of futures (those along the vertical axis), *Big+Deter-Mixed* has high success regret but zero or small days to completion regret. In these futures, the strategy fails more campaigns than alternative strategies, but for those campaigns it does complete, it does so as quickly as do the alternative strategies. In a relatively larger number of futures, *Big+Deter-Mixed* has high days to completion regret but zero success regret. In these futures, the strategy completes as many campaigns as the alternative strategies but can take longer to finish them. In those futures where *Big+Deter-Mixed* has high success regret, *Big-Cost* is always the strategy with the highest success rate. In futures where *Big+Deter-Mixed* has high days to completion regret, *Big+Deter-Time* (and never *Big-Cost*) often completes campaigns the fastest.

Figure 3.5 also notes two specific futures, labeled Future 1016 and Future 1152,[12] which we will now consider in more detail. Such detailed "drill-downs" into particular case studies are useful for three reasons. First, such case studies provide a way to validate outliers. A detailed analysis of an individual case helps to validate that the models are working as intended. Second, case studies provide a greater understanding of how the strategies actually perform across a 20-year future. Third, the case studies provide deeper insight into the tradeoffs among the strategies and why these tradeoffs occur.

---

[11] The goal of the scenario discovery algorithms is to find a parsimonious definition of a scenario that best captures futures of interest (i.e., futures where *Big+Deter-Mixed* has a success rate of at least 90 percent) and excludes all other futures (i.e., futures where *Big+Deter-Mixed* has a success rate of less than 90 percent). The "coverage" of the Moderate Scenario is 82 percent, i.e., 82 percent of futures in which *Big+Deter-Mixed* has a success rate of at least 90 percent are included in the Moderate Scenario. The "density" of the Moderate Scenario is 83 percent; i.e., *Big+Deter-Mixed* has a success rate of at least 90 percent in 83 percent of the futures included in the scenario.

[12] The numbers 1016 and 1152 refer to the order in which the futures were calculated during our runs.

Figure 3.5
Days to Completion Regret, Success Regret, and Success Rates for the
***Big+Deter-Mixed* Strategy for Each of the 361 Futures in the**
Moderate Scenario

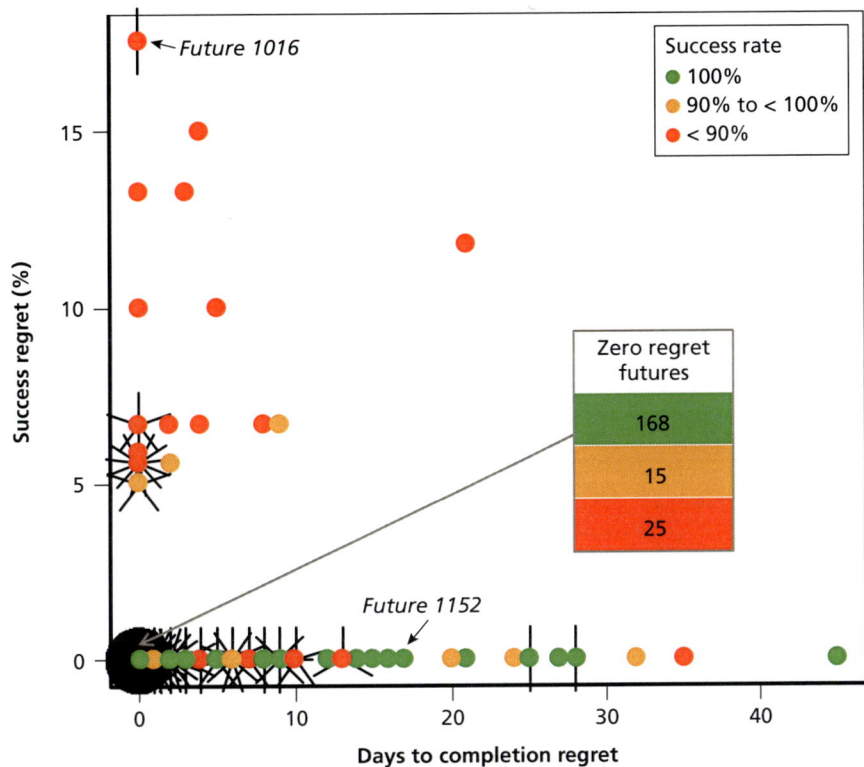

NOTES: Days to completion regret is calculated by comparing *Big+Deter-Mixed* days to completion against the days to completion of other strategies with the same or better success rate. For overlapping points, the whiskers show the number of overlapping points. No regret cases shows the distribution of points at the origin (zero days to completion regret and zero success regret).
RAND *RR1112-3.5*

In Future 1016 (labeled in the upper-left of Figure 3.5) the *Big+Deter-Mixed* strategy has its maximum regret; it fails in 18 percent of the future's campaigns, whereas the *Big-Cost* strategy successfully completes all the campaigns.[13] Future 1016 has security environment 21, which mirrors the past 20 years. The upper panel in Figure 3.6 compares the performance of the *Big+Deter-Mixed* and *Big-Cost* strategies over the progression of security conditions in this future. This security environment is characterized by several periods of relatively intense combat; *Big+Deter-Mixed* fails to complete combat in two of these periods. For example, the 2022 period presents security condition E (multiple major regional) with two major regional campaigns. The model simulates each campaign 100 times. *Big-Cost* is successful in all 200, whereas *Big+Deter-Mixed* is successful in only half these campaigns.

---

[13] Future 1048 overlaps Future 1016 in Figure 3.5. It has the same security environment as Future 1016 and has similar uncertainties; thus the behavior and results of the two futures are very similar.

Figure 3.6
Case Study of Future 1016 Comparing *Big+Deter-Mixed* and *Big-Cost* Strategies

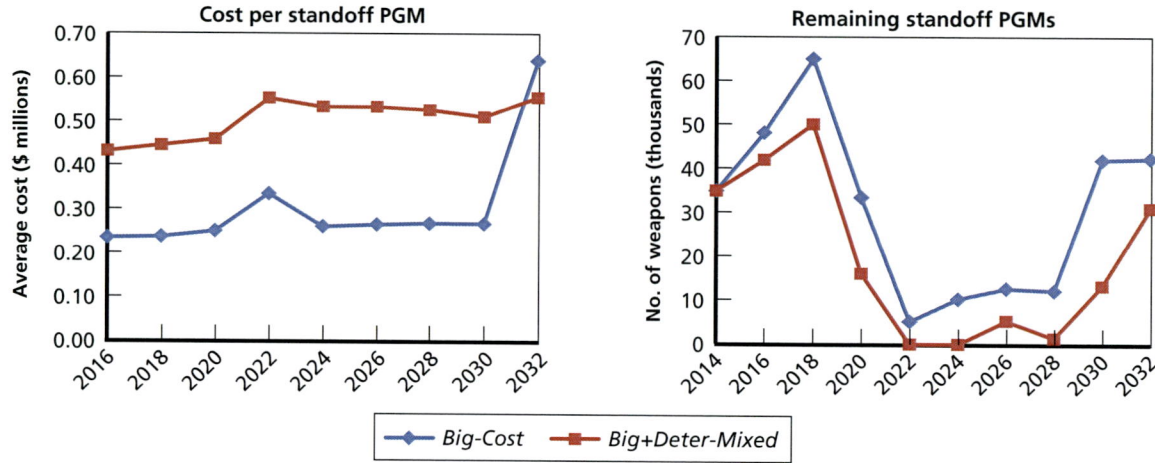

	2014	2016	2018	2020	2022	2024	2026	2028	2030	2032
Order	A	A	B	D	E	D	D	D	C	B
Targets	1,110	1,110	1,540	23,953	41,538	23,953	23,953	23,953	6,000	1,540
Big-Cost	100%	100%	100%	100%	100%	100%	100%	100%	100%	100%
Big+Deter-Mixed	100%	100%	100%	100%	50%	50%	100%	97%	100%	100%

NOTES: Percentages in the table on the top of the figure indicate the percentage of simulated campaigns that were completed successfully. Cost per standoff PGM is the cost of weapons acquired at the beginning of each period (no acquisition in 2014). Remaining standoff PGMs is the number remaining at the end of each period.
RAND RR1112-3.6

The lower panels of Figure 3.6 show some of the details that explain why in this future *Big-Cost* succeeds but *Big-Deter-Mixed* fails.[14] The *Big-Cost* strategy usually purchases standoff PGMs that are about half the cost of the weapons purchased by *Big+Deter-Mixed* (see the bottom left of Figure 3.6; strategies that minimize time to completion tend to buy more expensive weapons). Both strategies have the same maximum funding available in each period, thus *Big-Cost* will tend to purchase about twice as many munitions. The bottom-right panel of Figure 3.6 shows that this purchase pattern can prove disastrous for *Big+Deter-Mixed*—it twice runs out of standoff PGMs completely and therefore is unable to successfully complete the campaign. In this future, *Big+Deter-Mixed* never has the funds or the time needed to build up its standoff PGM inventories to its goals. The intensity of conflict also prevents *Big-Cost* from reaching its goals in most periods, but in 2030 and 2032, it has managed to reach its goals, so its inventories of standoff PGMs level out.[15]

In Future 1152 (labeled near the bottom of Figure 3.5) *Big+Deter-Mixed* has zero success regret but high days to completion regret. This is a future in which *Big-Cost* fails in two campaigns in the 2032 period where *Big+Deter-Mixed* is successful. *Big-Cost* fails in these futures

---

[14] Keeping a record of the parameters and variables generated by the models enables an after-the-simulation analysis. For example, quantities used of each weapons type are tracked for each campaign. This recordkeeping has a cost—a large number of variables must be recorded. *Big+Deter-Mixed* generated about 37 MB of data. However, this is a small cost relative to the computation time that would be needed to reproduce the data.

[15] *Big-Cost* goes above its goals in 2016 and 2018 because there remain inventories of existing standoff PGMs that the strategy no longer purchases. These existing inventories are expended during the campaigns.

because its goals are too small to stock sufficient quantities of munitions. *Big+Deter-Mixed* has sufficient weapons to complete all campaigns in this future but does not complete them as quickly as several other strategies.[16]

Future 1152 has security environment 24, which starts off as benign but enters a period of deterrence that is broken by major regional campaign (i.e., two types of campaigns in each period). As shown in the top panel of Figure 3.7, both strategies succeed until the year 2032, during which *Big-Cost* fails whereas *Big+Deter-Mixed* is successful. The lower panels of Figure 3.7 explain why. *Big+Deter-Mixed* increases its inventories of munitions throughout most of the future; however, it never reaches its full portfolio goals because the benign security environments in the early years provide only modest levels of munitions funding. In contrast, *Big-Cost*'s portfolio goals are much less costly, so it reaches them in 2020 and thereafter replaces only those weapons used during campaigns. In 2030, both strategies' inventories similarly deplete their inventories of standoff PGMs, but this reduction is a relatively greater percentage for *Big+Cost*. In 2032, *Big+Cost* is able to return its inventories to its goals; however, the legacy

Figure 3.7
Case Study of Future 1152 Comparing *Big+Deter-Mixed* and *Big-Cost* Strategies

	2014	2016	2018	2020	2022	2024	2026	2028	2030	2032
Order	A	A	A	A	A	B	B	B	D	D
Targets	1,105	1,105	1,105	1,105	1,105	1,530	1,530	1,530	23,724	23,724
Big-Cost	100%	100%	100%	100%	100%	100%	100%	100%	100%	0%
Big+Deter-Mixed	100%	100%	100%	100%	100%	100%	100%	97%	100%	100%

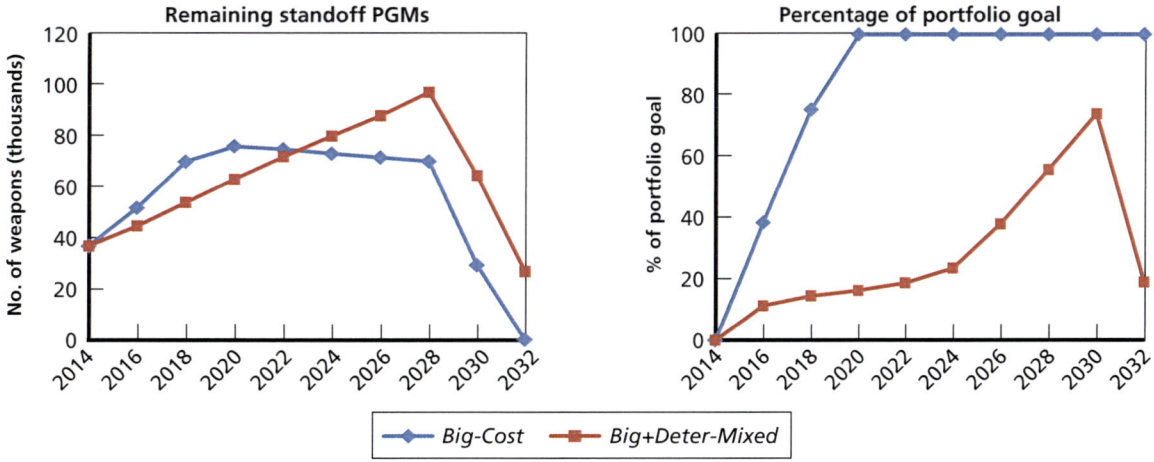

NOTES: Percentages in the table on the top of the figure indicate the percentage of simulated campaigns that were completed successfully. Remaining standoff PGMs is the number remaining at end of the each period. Percentage of portfolio goal is the minimum percentage across all munitions after acquisition at the beginning of the period.

RAND *RR1112-3.7*

---

[16] *Big+Deter-Mixed* completes the campaigns in 88 days on average, whereas *Big-Time* and *Big+Deter-Time* complete the campaigns in 73 days and *Big+Deter/GPS-Mixed* completes the campaigns in 71 days. Therefore, *Big+Deter-Mixed* has a days to completion regret of 17 days.

weapons remaining from the current baseline were mostly used in 2030,[17] and the inventories are no longer sufficient to complete either campaign in 2032.

Performance of the *Big+Deter-Mixed* Strategy in the Extreme Scenario

This study also compared the six alternative strategies in the Extreme Scenario, which includes the 889 futures in which *Big+Deter-Mixed* generally has a success rate less than 90 percent. Table 3.3 summarizes the results. As in the Moderate Scenario, *Big+Deter-Mixed* and *Big-Cost* are the two dominant strategies. The relative performance of *Big+Deter-Mixed* compared to the other strategies is even better in the Extreme than in the Moderate Scenario.

Overall, however, none of the alternative strategies considered in this analysis have sufficiently high success rates in the Extreme Scenario. Larger, more costly weapons portfolios would clearly increase success rates in this scenario. In addition, future RDM analyses might consider other policy options, such as developing a surge capability in the munitions industrial base or ensuring sufficient warning to increase munitions spending in advance of large campaigns. However, it is worth noting that the current analysis provides the valuable service of clearly delineating the types of futures beyond which the munitions mix strategies under consideration cease to perform well.

Future Focus on Purchase Rules in Addition to Portfolio Goals

Most munitions mix analyses, including this one, focus on what we call here portfolio goals, which specify the desired number of weapons of each type policymakers aim to purchase. But as this analysis makes clear, the purchase rules, which specify the order in which munitions

Table 3.3
Summary of Measures in the Extreme Scenario (889 Futures)

Strategy	Maximum Adjusted Cost Regret ($ Millions)	Maximum Success Regret (%)	Number with Zero Success Regret	Number with Zero Success and Cost Regret	Average Cost ($ Millions)	Average Success (%)
Baseline	43,924	44	148	73	57,277	60
Big-Cost	0	43	341	341	31,483	61
Big-Time	58,646	21	505	86	69,243	67
Big+Deter-Time	58,802	20	524	57	69,261	67
Big+Deter-Mixed	56,117	16	761	278	68,419	70
Big+Deter/GPS-Mixed	58,853	20	579	54	69,495	68

NOTE: The shaded rows indicate the two best-performing strategies.

---

[17] The *Big-Cost* strategy ran out of four legacy PGMs completely in 2030, which it does not purchase as part of its strategy: AGM-154 (Joint Standoff Weapon), AGM-158 (Joint Air-to-Surface Standoff Missile-ER [extended range]), BGM-109E (Tomahawk Land-Attack Missile), and MGM-168 ATACMS (Army Tactical Missile System).

will be replaced when the stockpile is depleted, may be at least as important in determining the success of any munitions mix strategy. In many futures, the *Big+Deter-Mixed* and *Big-Cost* strategies fail to complete campaigns, or fail to complete them quickly, because a series of earlier campaigns have depleted stockpiles, leaving them poorly matched to the needs of the next conflict.

Future RDM analyses could usefully focus on a wider and more carefully tailored set of alternative purchase rules than those considered here. However, the current analysis is sufficient to suggest some hypotheses regarding more successful purchasing rules. For instance, a step-wise purchasing strategy might prove promising. Such a strategy might initially, when faced with depleted inventories and limited funding, focus on purchasing economical weapons, similar to those in the *Big-Cost* strategy. Such purchases would aim at inventories able to successfully complete campaigns, even if it took a long time to do so. As inventories and funding increased, the strategy could next focus on purchasing more expensive weapons that can complete campaigns quickly. As a simple first step toward modeling such a step-wise strategy, one might design a purchase rule that first invests toward the *Big-Cost* portfolio, and—once achieved—switch to another portfolio that buys pricier weapons that aim to minimize time. Such a strategy should, at a minimum, perform at least as well as the *Big-Cost* strategy. It would also make clear the tradeoffs between spending additional resources on more effective weapons in the present and the risk of spending additional resources fighting longer campaigns in the future.

Note that such a step-wise strategy would focus on achieving robustness adaptively, that is, successfully evolving over time in response to new information. It would also exemplify several of Danzig's recommendations for preparing for predictive failure. The strategy would delay some decisions until more funding is available, prioritize cost-effective standoff weapons that are adaptable to a variety of future contingences, and purchase more for the short term. Such a step-wise strategy would first invest to ensure success and then invest to reduce time, total costs of campaigns (including the costs from lengthening campaigns), and risk.

CHAPTER FOUR

Conclusions

This initial application to the munitions mix challenge suggests that RDM can provide useful inputs to defense planning. The approach provides a systematic, detailed, quantitative means to plan defense investments while acknowledging the likelihood of predictive failure. RDM stress-tests plans over a wide range of plausible futures, suggests scenarios that illuminate the distinguishing characteristics of futures in which strategies do and do not meet their goals, and helps decision makers use this information to develop more robust plans and evaluate the tradeoffs among them. RDM can help identify and evaluate adaptive strategies, ones designed to evolve over time in response to new information. Danzig (2011), among others, faults DoD for an overreliance on prediction in its planning. He recommends robust and flexible strategies as a response to the likelihood of predictive failure. Although RDM is not a substitute for DoD's current reliance on scenario-based analysis, with its overconstraining set of assumptions, this research project suggests how this new approach could complement traditional analysis. Such synergy could not only provide DoD with a significantly broader set of alternative futures to evaluate but also enable analytical economy by focusing these more costly techniques on scenarios of highest interest. This initial application suggests that RDM could not only help DoD conduct the type of planning demonstrated here—evaluating and implementing resource-focused strategies—but also conduct programmatic-, policy-, and operational-related analysis associated with PPBS,[1] planning scenarios, and courses of action.

A Robust Munitions Mix Strategy

The findings in this report emerge from simplified models and unclassified data, so at best are suggestive rather than definitive. Nonetheless, this analysis finds that a munitions mix strategy that we call *Big+Deter-Mixed* is robust over a wide range of plausible futures. The following summary of this strategy and its performance may provide useful insights for current munitions mix decisions and suggest more generally the types of information available from an RDM analysis.

The *Big+Deter-Mixed* strategy consists of a portfolio goal, which specifies the desired number of weapons of each type, and a purchase rule, which specifies the order in which munitions will be replaced when the stockpile is depleted. We constructed the portfolio goal by using a WoT optimization model to give the weapons mix that provides the best balance

---

[1] PPBS is DoD's major decision making process, composed of planning, programming, and budgeting system. DonVito (1969).

between weapons' acquisition cost and time to completion for two planning scenarios: (1) a deterrence campaign with a small number of targets accessible only to standoff weapons and (2) two back-to-back medium-size campaigns. We chose this set of planning scenarios through an iterative process of stress-testing strategies with portfolio goals derived from alternative sets of planning scenarios. *Big+Deter-Mixed* uses a purchase rule we call Replenishment, which restocks weapons inventories in proportion to shortages in the inventories. We considered two alternative purchase rules and chose replenishment as the superior one.

We stress-tested *Big+Deter-Mixed* and five alternative strategies over a wide range of futures that combine assumptions about both large-scale factors—alternative security environment with varying levels of severity—and small-scale factors—alternative values for parameters representing weapons effectiveness, adversary capabilities, and tactical decisions. We evaluate the strategies with three measures of performance: their ability to complete campaigns (success rate), the speed with which they complete campaigns (days to completion), and the total cost of acquiring and replenishing the weapons portfolio (cost).

A scenario discovery statistical cluster analysis, applied to the model-generated database of thousands of futures, identified two important scenarios around which we organized our comparisons of the strategies. The Moderate Scenario contains those futures in which *Big+Deter-Mixed* has a generally high success rate (greater than 90 percent) and the Extreme Scenario contains those futures where *Big+Deter-Mixed* has a generally low success rate (less than 90 percent). The most important uncertainties distinguishing these two scenarios are the severity of the security environment and the effectiveness of GPS weapons. These two scenarios—Moderate and Extreme—suggest that with a high level of GPS effectiveness and in security environments up to the severity level of the last 20 years, *Big+Deter-Mixed* will have a generally high success rate over a wide range of assumptions about other uncertainties. If GPS effectiveness is low, the *Big+Deter-Mixed* strategy will have high success rates only in security environments about half as severe as those in the past 20 years.

Of the six strategies considered in this analysis, *Big+Deter-Mixed* is the most robust in both scenarios, in the sense that it performs better than the alternatives for each of the three measures—success rate, completion time, and cost—over a wide range of futures. A strategy called *Big-Cost* costs about a fourth of *Big+Deter-Mixed* in all futures, but in most futures (but not all), it has lower success rates and substantially longer completion times.

Big+Deter-Mixed does have vulnerabilities, however. In the Moderate Scenario, *Big+Deter-Mixed* completes campaigns in some futures significantly more slowly than in a strategy we call *Big-Time*, which costs about the same but stocks a larger number of expensive standoff weapons. In these futures, *Big-Time*'s extra, high-performance PGMs make a critical difference. Somewhat surprisingly, *Big+Deter-Mixed* also has lower success rates in some futures than *Big-Cost*, which stocks primarily less-expensive weapons. In these futures, in which the United States fights a series of closely spaced conflicts, the replenishment purchase rule focuses too many resources on restocking expensive PGMs, so that *Big+Deter-Mixed*'s munitions portfolio is misaligned with the needs of subsequent campaigns.

This vulnerability analysis suggests ways to adjust *Big+Deter-Mixed*'s purchasing rule that might eliminate this vulnerability with respect to the *Big-Cost* strategy. However, the vulnerability analysis does not suggest ways to adjust *Big+Deter-Mixed* (other than spending more on munitions) to eliminate the vulnerability with respect to the *Big-Time* strategy.

In the Extreme Scenario, *Big+Deter-Mixed* generally performs better than the alternative strategies but has insufficiently high success rates, because the scenario's campaigns require far more weapons than contained in the stockpiles. To reduce these vulnerabilities, the United States could spend more on munitions or consider other policy options, such as developing a surge capability in the munitions industrial base or ensuring sufficient warning time to increase munitions spending well in advance of large campaigns.

The Future of RDM in Defense Planning

This initial application of RDM to the munitions mix challenge provides a proof of concept showing how RDM might be employed for defense planning, which was the research project's overarching goal. This particular munitions mix application demonstrates that RDM can provide the types of analytic information DoD might find useful in identifying robust and flexible strategies that can achieve success despite predictive failure.

Future RDM defense planning applications would need to address a number of challenges.

First, to the extent that future RDM applications used models more complicated than those used in this study, they might need to draw on more powerful computational resources than we had available. The analysis here used relatively simple models with no classified data but still faced computational constraints. In particular, using complex models could take several weeks to conduct each iteration of the analysis, which reduces the rate at which we could design and explore new strategies.[2] DoD has access to more capable computational platforms, such as the defense computing resources (e.g., those of DoD's High Performance Computing Modernization Program). Using more powerful computational resources could have two major benefits. First, it would shorten the time between developing a new strategy and conducting the model runs necessary to assess that strategy. Thus, a greater number of iterations could be run in a given amount of calendar time to develop strategies that are robust to vulnerabilities discovered in each iteration. Beyond just adding more iterations, it would allow the research team to interactively investigate more policy options and optimization approaches. Second, it would allow the models to assess a larger, more realistic variety of targets, weapons, delivery platforms, and operational and environmental conditions. However, as RDM is integrated into defense planning, the computational requirements may also increase as the models used become more complicated.

Second, future RDM applications might need to employ a wider set of models, beyond the types presently used in current DoD planning exercises. For instance, the importance of purchasing rules in the Moderate Scenario, and the demands of the Extreme Scenario, suggest that in seeking robust strategies, an RDM munitions mix analysis might usefully consider policies that affect the munitions industrial base and logistics system, in addition to WoT campaign models. As an example, a more comprehensive follow-on analysis might consider the effect of weapons and manufacturing technology, acquisition and testing, weapons industrial base opportunities and limitations, economies of scale in both purchase and delivery of munitions, global redistribution of munitions, and integrated logistics during periods of conflict. Traditionally, the MRP employs only campaign models, so a more complete RDM analysis

---

[2] Running one strategy over 1,250 futures took about a week under ideal conditions (i.e., a dedicated core of a processor was available to process the strategy). However, computation time was much longer when the server was busy with other RAND projects.

could call for a significant expansion of the elements of the full system considered in DoD's munitions mix studies.

Finally, integrating RDM into defense planning could raise a number of analytical opportunities. Initially at least, and perhaps as a permanent approach, RDM might be used as a precursor to, or in parallel with, traditional processes. For instance, RDM could stress-test munitions mix strategies generated by the current MRP over a much wider range of futures and suggest new scenarios in which these strategies might be considered. In addition, RDM could provide guidelines for potential robust strategies, which could then be tested and fleshed out using traditional tools for defense planning. Such integration would raise important questions. For instance, RDM analyses that used more complicated models would involve and produce vast amounts of data—requiring means to make them transparent, traceable, and credible. Alternatively, RDM analyses with very simple models might serve as useful screening tools to situate more traditional, detailed analyses. Future research and experience might help clarify the situations in which RDM analyses might use simpler or more complicated models as part of overall, integrated planning processes.

Above and beyond these considerations of conducting and integrating different types of analysis, RDM might raise an issue of process and communication for DoD. For instance, RDM often stress-tests strategies until they break. This provides useful information, but DoD might have to employ security-sensitive procedures to internally and externally manage and communicate information regarding the vulnerabilities of their proposed policies.

We recommend that RDM be used to initially supplement current defense planning. As RDM continues to validate itself in the national security analytical realm, as it has in infrastructure planning, it can be better integrated into PPBS and deliberative planning activities. Shifting the defense planning processes to a more balanced approach, through the integration of RDM, would involve issues of adjustment, but we expect that RDM's benefits will far exceed its costs—to the benefit of the Pentagon, the Congress, and the nation. RDM would improve defense planning by enabling DoD to examine its strategies, policies, plans, programs, and budgets over a wide range of futures, identify vulnerabilities, and design responses that reduce those vulnerabilities. This in turn would improve DoD's ability to design and evaluate robust and flexible strategies. RDM can help DoD more successfully achieve its goals in a world in which surprise and uncertainty are virtually certain to lead to predictive failure.

The Weapons on Target Model

Introduction

This appendix overviews the WoT model and then describes WoT in functional terms, its key techniques and algorithms, its limitations and artificialities, its coding standards, its modes of operation, and its data file.

Overview

WoT is a time-stepped Monte Carlo simulation of strike campaigns. It attempts to quickly achieve a user-specified objective (the destruction of a given fraction of a specified target set) within the limits of weapon inventories, weapon delivery rates, and the anti-access capabilities of adversaries. It does so using user-specified alternative strategies against high-priority targets and limits on weapon use such as constraints on collateral damage.

Many tradeoffs are possible in the design of force-on-force models such as WoT. Speed and detail can be traded off against each other, as can transparency and validity. The RDM process requires a fast and transparent model (because of the large number of cases to be treated and because of the need to explain results). To illustrate its speed, when run on a laptop computer with trials entailing 6,280 targets having 25 priorities, WoT completed 100 trials in 57 seconds. WoT provides a high level of transparency by relying on a few simple techniques and algorithms. It also features over a dozen "probes" that enable users to follow and understand its operation. For example, one such probe allows users to follow day-by-day progress as campaigns proceed. Another probe illuminates the weapon selection process, and so on.

WoT adheres to certain user-imposed rules and constraints. For example, users may impose a shoot-look-shoot strategy or a shoot-shoot-look strategy against high-value targets. As another example, users can constrain weapon use (perhaps by prohibiting the use of large "dumb" bombs against targets in urban areas). WoT has three modes of operation: normal, time-minimizing, and cost-minimizing. In its normal mode of operation, WoT seeks to kill prioritized targets as effectively as possible within the limits of weapon inventories, weapon delivery rates, and the previously described user-imposed rules and constraints. In its time-minimizing mode, WoT identifies the weapon mix yielding the most rapid achievement of objectives subject to weapon delivery rate limits. In its cost-minimizing mode, WoT identifies the munitions mix yielding the lowest procurement-cost munitions mix with which objectives can be achieved.

WoT imposes minimal constraints on users. In particular, it has no meaningful limit on the number of targets or target types, weapons or weapon types, means of weapon delivery, and so on.

WoT output includes the following:

- summary material describing scenarios, the mode of WoT operation employed, and the weaponeering tactic selected against high-priority targets
- a description of weapon delivery rate capabilities, in terms of surge and sustained rates
- a summary of the level of success achieved, including the overall average number of targets killed and the fraction of targets killed; WoT's objective, also stated as a percentage of targets killed, is also provided
- weapon expenditures by weapon type, expressed as the mean and standard deviation of the number of weapons expended across all trials
- weapon costs by weapon type, expressed as the mean and standard deviation of the procurement cost of weapons expended across all trials; total weapon procurement cost is also provided
- additional summary information, such as the cumulative probability of successful termination by day
- information generated during trials to facilitate testing and improve understanding of outcomes, including day-by-day activity and result summaries, progress in achieving user-specified subobjectives, planning processes, and weapon selection processes
- outputs from "probes" described previously.

Techniques and Algorithms

Two key techniques in making WoT run efficiently are prioritization of targets in the WoT data set and the preprocessing of Pks. Target prioritization enables the efficient target selection process shown in Figure A.1. Preprocessing Pks enables the simple weapon selection process shown in Figure A.2.

The so-called "greedy algorithm" and the Mersenne Twister are two key algorithms in WoT. Greedy algorithms follow the problem-solving heuristic of making the locally optimal choice at each stage with the hope of finding a global optimum. WoT uses the greedy algorithm for the prioritized target set in all modes of operation.[1] Although greedy algorithms are generally very efficient means for finding optimal solution, they do not always do so. Sometimes they yield locally optimal solutions that approximate a global optimal solution in a reasonable time. The problem of mistaking a locally optimal solution (a "bump") for a global optimal solution (a "peak") was not seen in the WoT application when optimal solutions were found manually. More abstractly, the appropriateness of the greedy algorithm in minimizing cost or campaign time is illustrated by a simple thought experiment based on a simple question: Can the addition of an option (not forced) to use a weapon increase the minimum time or cost to complete a campaign? Such an option clearly cannot force an increase in the cost or time to complete a campaign because the option can be ignored. In the WoT application, the greedy algorithm resembles the steepest-edge method used in linear programming. Many organiza-

---

[1]   In its normal and time-minimizing modes of operation, WoT selects the usable weapon with the highest Pk against any given target. In its cost-minimizing mode, WoT selects the usable weapon with the highest ratio of Pk to unit cost (the weapon providing the most "bang for the buck").

Figure A.1
Target Selection in WoT

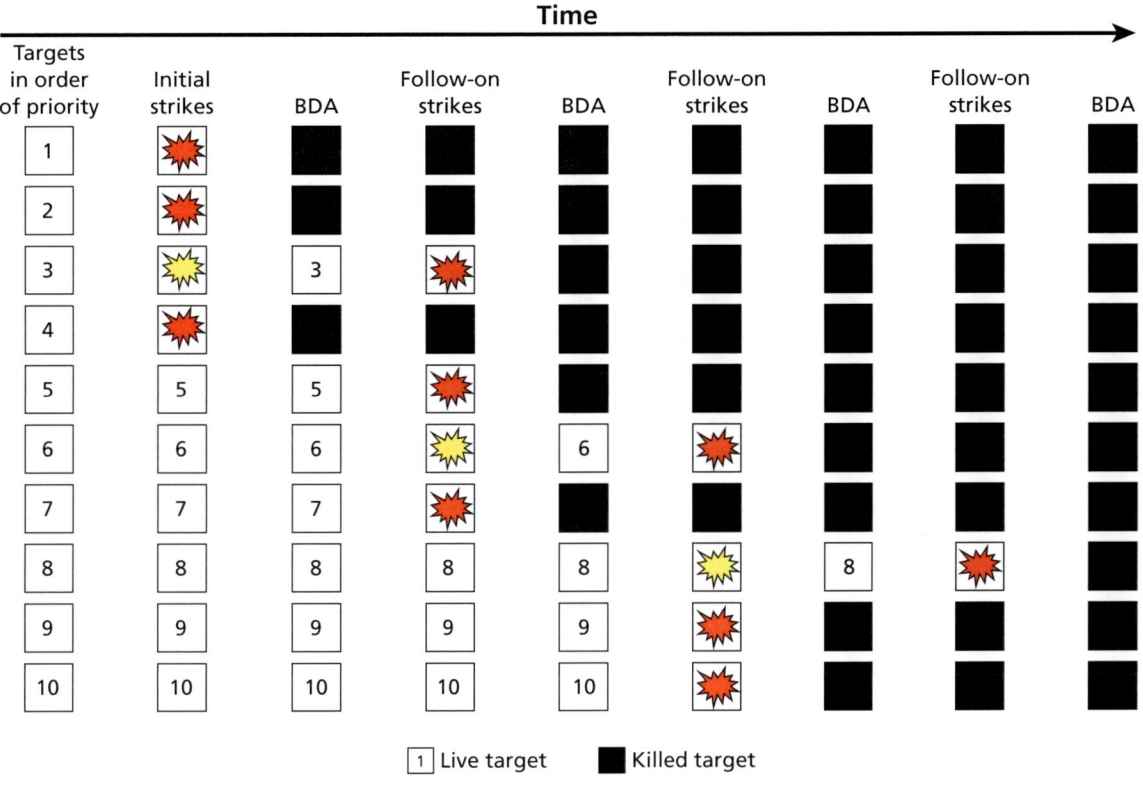

RAND RR1112-A.1

tions, for instance the assessment branch of the Office of the Chief of Naval Operations, N-81, find the greedy algorithm an effective approach in their campaign analyses.

A high-quality pseudorandom number generator is critical to generating valid results in Monte Carlo simulations. The Mersenne Twister is currently regarded the current gold standard of pseudorandom number generators for Monte Carlo simulations.[2] To illustrate, the period of a pseudorandom number generator is a simple indicator of its quality. The Mersenne Twister algorithm has a staggering period of 1.3×10^{90} samples. The Mersenne Twister also has the advantage of portability; results generated using a 32-bit Windows computer are identical to those generated on a 64-bit Apple Macintosh computer.

Limitations and Artificialities

Fleeting and dispersed targets were added late in the development of WoT. Testing against such targets using probes indicated no operational problems. However, problems were reported when optimizing weapon inventories against target sets including such targets. Project resource limitations prevented exploration of these reports, so WoT should not be used in its optimization modes against target sets including fleeting or dispersed targets.

---

2   See University of Michigan (undated).

Figure A.2
Weapon and Target Pairing in WoT

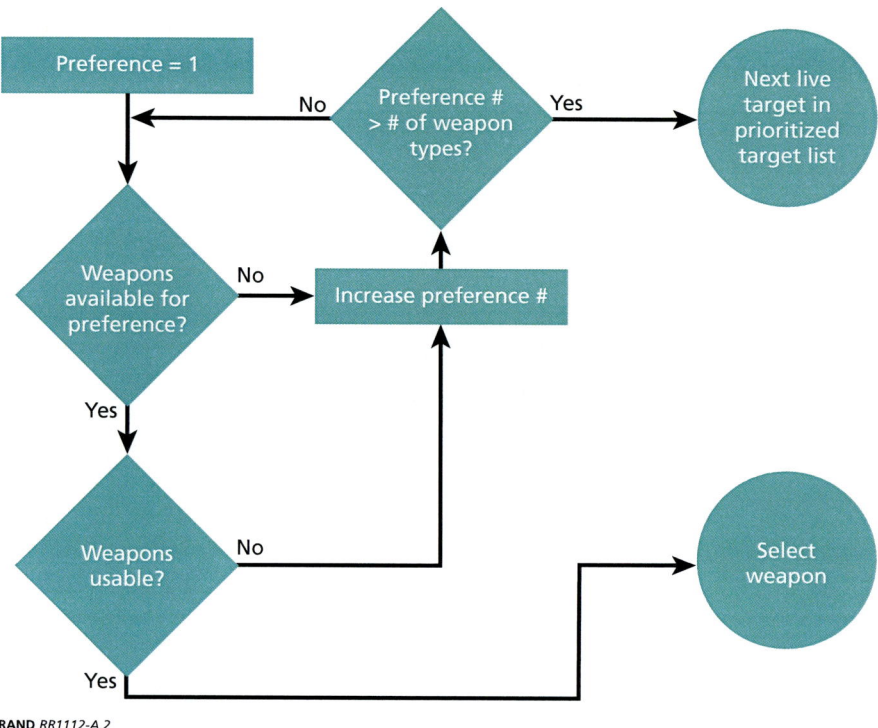

RAND *RR1112-A.2*

In its cost-minimizing mode, WoT identifies the weapon mix with the least procurement cost capable of meeting scenario requirements. This process was recognized as suboptimization; it would be better to identify the weapon mix with the lowest total warfighting cost capable of meeting scenario requirements. A process for identifying such a weapon mix was developed, but project resource limitations prevented its implementation.

The greatest artificiality of WoT is in its treatment of adversary air defense systems. Rather than rolling back air defenses and progressively exposing targets to attack, WoT treats air defenses in an all-or-nothing manner. In principle, such treatment could be avoided by simply associating targets with the defense systems protecting them and allowing the target to be struck freely when all such defensive systems have been killed. However, such detailed modeling lies beyond the scope of this project. The effects of this artificiality can be reduced by describing the expected air defenses that must be killed to achieve a semi-permissive environment.

Lack of reconstitution capabilities for Red is another artificiality in WoT. Reconstitution capabilities could be added with data describing reconstitution rates.

WoT does not treat Blue attrition. Blue losses are not counted, and sustained weapon delivery capacities do not degrade over the course of conflicts. Implementing Blue losses was determined to be beyond the scope of this project, but a placeholder method that could be used to implement Blue losses was built into WoT.

Perfect bomb damage assessment (BDA) is the last noted artificiality in WoT. Targets are determined to have been killed if and only if they have actually been killed. The simulation of imperfect BDA is thorny both for failure to recognize that a target has been killed or the false belief that a target has been killed. At present, the best way to incorporate the effects of imperfect BDA is a "tax" on Pks to increase weapon expenditures.

Functional Description

WoT is described here functionally in terms of overall organization, initialization, operation, and termination.

Organization

WoT is organized simply using just three methods (essentially subroutines):

1. a method for starting up the model
2. a method for executing trials, generating intermediate results, and outputting final results
3. a method for shutting down the model.

WoT Initialization

Initialization of WoT on startup begins by priming a pseudorandom number generator with fixed seeds. This assures that its results are reproducible. WoT then opens the input file "WoT.dat" determining the name for an output file to be used by Analytica, the name of the scenario (which can include material used by Analytica), and possibly other files used for testing or special purposes. One such special purpose is determining the weapon mix with minimum procurement cost—here WoT opens the file "OptWpns.dat" into which it will write that optimum weapon mix. Another special purpose is enabling the user to rerun cases of interest—here WoT opens the file "seed.log" so that a trial of interest can be recreated using the original seeds for the pseudorandom number generator to reproduce and examine previous results.

Operation

Operation begins with trial setup, preprocessing weapon effectiveness data, trial initialization, daily operations, and trial termination.

Trial Setup

Trial execution begins by setting up the scenario for the trial. This entails the following steps:

1. Determine the initial permissiveness and the requirement to achieve a permissive environment in the event that the initial environment is nonpermissive.
2. Determine the maximum number of days a scenario simulation will be allowed to run and the number of Monte Carlo trials to be conducted.
3. Determine a weapon rule to be used in deciding on the number of weapons to be allocated against high-priority targets by level of permissiveness. The choices are (a) use a single weapon, (b) use a pair of weapons to more quickly defeat air defenses, and (c) use as many weapons as needed to achieve 95 percent confidence that air defense targets will be killed.
4. Determine weapon groupings. The weapon group construct is used to capture the idea of competition for delivery resources. It also captures the idea of independence among delivery resources. For example, there is competition among gravity weapons

for aircraft. On the other hand, the number of surface-to-surface weapons that can be launched in a given day is independent of the number of strike sorties that can be generated on that day

5. Set surge and sustained rates, which are described by weapon group using the number of surge days possible, the surge delivery rate, and the sustained delivery rate. For some weapon systems, such as Tomahawk land attack missiles, surge rate is not applicable. Surge and sustained rates are set equal for such weapon systems.

6. Set the number of distinct types of weapons, the number of weapons of each type available at the outset of the scenario, the permissiveness condition under which a given weapon type is usable, and the weapon group to which an individual weapon belongs.

7. Define holders for intermediate and end results. For example, holders for the mean and standard deviation of weapons expended are defined by weapon type.

8. Define the terms needed for preprocessing data (described below).

9. Define the terms needed to characterize individual targets. For example, targets are characterized by priority, as live or killed, distributed, and so on.

10. Read user-defined Pks for each weapon against each target type as well as the unit cost of each weapon.[3]

11. Compute the cost-effectiveness of each weapon against each target type. Cost-effectiveness against a given target type is computed as Pk divided by unit cost. Weapon cost-effectiveness will be used only in WoT's cost-minimization mode.

12. Read target priorities by type and assign priorities to individual targets.

Preprocessing Weapon Effectiveness Data

Following initial setup of the scenario, WoT preprocesses inputs to speed up weapon selection. In its normal or time-minimizing modes, WoT attempts to use weapons with the highest Pk against each target. In its cost-minimizing mode, WoT attempts to use the most cost-effective weapon against each target. Pk values and cost-effectiveness scores are sorted accordingly. This gives WoT a prioritized list of weapons for use against each target type according to its mode of operation.

Trial Initialization

Following preprocessing, metrics that crosscut trials are initialized. For example, total weapon procurement cost is initialized to 0; it is incremented as trials are completed.

Daily Operations

WoT makes certain determinations at the outset of each day. Have any scheduled replenishments arrived? Are any surviving fleeting targets exposed? If a fleeting target has just become vulnerable, how long will it remain so? With these determinations, a strike plan is generated and strikes are conducted accordingly. At the end of each day of operation, BDA is conducted to determine stochastically which engaged targets were killed in the previous day and which targets survived. After completing BDA, a status assessment is performed. Have priorities been achieved? If initial conditions were nonpermissive, are conditions now permissive?

---

[3] Users can assign Pks to prevent inappropriate weapon use. For example, a weapon that would create unacceptable collateral damage against a given target type can be assigned a Pk of 0 against that target type. As another example, a weapon that represents overkill against a given target could in turn be assigned a Pk of 0 against that target type.

WoT, in its normal mode of operation, begins with the highest-priority targets and assigns the highest Pk usable weapon against each of them. A weapon is declared usable if (1) there is delivery capacity for it, (2) the weapon can be used in the current permissiveness environment, and (3) WoT has not been forbidden to use the weapon against the target.[4] Strikes are conducted daily with each day's strikes followed by BDA. The process of striking targets by priority is illustrated in Figure A.1, which illustrates the simple case of ten targets and a weapon delivery rate limited to four weapons per day.

WoT normally selects weapons according to their probability of kill. Choices are made within the constraints of weapon inventories, usability, and the availability of means of delivery. This process is illustrated in Figure A.2. WoT first decides if a preferred weapon is available. If it is not, it moves on to its next-preferred weapon. If a weapon is available, WoT tests for its usability. As noted previously, an available weapon is declared usable if (1) there is delivery capacity for it, (2) the weapon can be used in the current permissiveness environment, and (3) WoT has not been forbidden to use the weapon against the target. Hence, weapon selection is done by order of preference against the target of interest with the requirement that there is at least one weapon remaining in the portfolio, that there is means to deliver the weapon, that the weapon can be used in the current permissiveness environment, and that the use of the weapon against the target of interest is acceptable to the user.

Trial Termination

Trials are terminated when user-specified objectives have been achieved or when WoT determines that those objectives are unachievable. As an example of the latter determination, WoT will terminate when all weapon inventories have been exhausted or when all remaining weapons are unusable against a live target that must be killed to meet user-specified requirements.

Run Termination

WoT executes a user-specified number of trials for each case. WoT prepares and outputs final results when those trials have been completed.

WoT Data Files

A short sample WoT data file is provided in Figure A.3. Line numbers in the example have been added for clarity; line numbers are not present in actual data files.

A line-by-line description of the data file is now provided to explain WoT setup and to provide insights into the operation of WoT.

Line 1. This is a scenario description. It can be arbitrarily long. For this sample data file, WoT would designate the scenario "Sample Short Data File."

Line 2. This is a 0/1 Boolean entry indicating whether the scenario is initially considered permissive. Low-level conflicts are examples of initially permissive scenarios. Conflicts with adversaries having strong anti-access/area denial capabilities are examples of initially nonpermissive scenarios. The sample data file indicates an initially nonpermissive environment.

---

[4] For example, weapon use against certain types of targets can be forbidden because their use would cause unacceptable collateral damage or because they represent "overkill." Setting the weapon's Pk against the target to 0 signals forbidden use.

Figure A.3
A Sample WoT Data File

```
 1        Sample Short Data File
 2        1
 3        1
 4        100
 5        100.0
 6        200
 7        1
 8        2
 9        0            2           100        50
10             1       999        100        100
11             2
12             85      0.130   0          0
13             10      1.447   0          0
14             999
15             10
16             10
17             4
18             4  1    0  0        0
19             8  2    0  0        0
20             12 3    0  0        8
21             16 4    0  0        0
22             0       0.33 0.50
23             1       0.52 0.50
24             2       0.31 0.35
25             3       0.34 0.34
```

Line 3. This indicates the target priority level for achieving permissive conditions. In this example, only the highest-priority targets must be killed to achieve permissive conditions.

Line 4. This indicates the maximum number of days (here 100) that trials will be allowed to run. In addition to allowing users to impose a time limit on campaigns, this input serves as a guard against runaway trials that WoT for some reason cannot terminate. Such trials were eliminated in testing.

Line 5. This indicates the required percentage of targets to be killed for success. It is used, as will be discussed below, to temper the effects of all-or-nothing anti-access capabilities for the adversary.

Line 6. This is the number of trials to be conducted.

Line 7. This specifies a strategy for assigning weapons against high-priority (anti-access system) targets. Here, the entry 1 indicates the use of a shoot-look-shoot strategy with high-priority targets struck using a single weapon followed by BDA. The entry 2 indicates a shoot-shoot-look strategy in which high-priority targets are struck with two weapons before conducting BDA. This strategy accelerates the achievement of permissive conditions at the cost of increased use of the most capable weapons. The entry 3 indicates a strategy against difficult-to-kill high-priority targets in which weapons are assigned until a 95 percent probability of kill is achieved. This strategy can accelerate the achievement of permissive conditions more than the shoot-shoot-look strategy—at a higher cost in the use of the most capable weapons.

Line 8. This is the number of weapon groups to be used. A simple weapon grouping might use two weapon groups: surface-to-surface missiles and air-delivered weapons. A more sophisticated grouping might organize weapons into surface-to-surface missiles, those delivered by bombers, and those delivered by fighter/attack aircraft. Without weapon groups (i.e., using an all-encompassing weapon delivery rate), we would have such anomalies as the inability to launch a JDAM because the weapon delivery rate was exhausted by TLAMs. Put another way, weapon groups allow for independence in the use of weapons with differing means of delivery.

Lines 9 and 10. Weapon groups are characterized by weapon delivery rates in surge and sustained operations. These lines indicate the duration of surge operations and the number of weapons per day in surge and sustained operations. In this example, there are two weapon groups with the first group able to surge for two days with a surge rate of 100 weapons per day followed by a sustained rate of 50 weapons per day. The second group in this example is able to deliver 100 weapons per day under all conditions. It might reflect surface-to-surface missiles, to which the surge concept does not apply.

Line 11. This is simply the number of types of weapons. In this simple example, there are only two types of weapons.

Lines 12 and 13. These lines describe the two types of weapons specified above. The initial number of weapons in the portfolio, weapon unit cost, usability in nonpermissive environments, and the weapon group associated with the weapon are specified.

Line 14. This is the frequency with which resupply is conducted. The value 999 entered into data assures that there will be no resupply in campaigns that cannot extend beyond 100 days.

Lines 15 and 16. These lines reflect the number of weapons of both types added to the portfolio when replenishment occurs (ignoring the fact that with replenishment scheduled every 999 days, there will be no replenishment).

Line 17. This line identifies the number of distinct target types in the target set.

Lines 18–21. Each of these lines characterizes the targets of a given type in the target set. The first entry specifies the number of targets of the specified type. The second entry specifies the priority of the target type. The next two entries are used to characterize fleeting and distributed targets. Fleeting targets are characterized by the mean time between periods of exposure and the duration of periods of exposure. Distributed targets are characterized by their number of designated mean points of impact (DMPIs): Point targets can be identified as having 0 or 1 DMPI. To illustrate, line 20 ("12 3 0 0   8") indicates that there are 12 targets of this type, they are third in priority, they are not fleeting, but are dispersed with 8 DMPIs.

Lines 22–25. The last lines of the WoT data set specify weapon Pks for target/weapon pairs. Thought of as a table, each row reflects a target type and each column reflects a weapon type. As noted above, weapon use against a set target type can be prohibited by setting the associated Pk to 0.

The WoT Header File

WoT uses a single so-called header file, "WoT.h," to define the objects (such as weapons and targets) in WoT. The header file is used to set the mode of operation, with instructions for doing so in the header file. The header file can also be used to activate and deactivate probes that can be used for testing and increasing transparency.

Coding Standard

WoT is written in the 2003 version of ANSI-standard C++. The 2003 standard was selected for portability; it was thought possible that some users' systems might not have the most recent (2011) C++ compiler. Also, little coding advantage was seen in the most recent standard for this application. DoD coding standards were used to develop WoT.

WoT has eight points of internal self-testing. For example, WoT checks to see if targets have been entered in priority order. If not, an error message indicates the problem and its location. WoT always terminates when it detects an error so that such errors cannot be ignored. WoT also uses a belt-and-suspenders approach in some areas with results computed by two separate means using different inputs. Failure to agree between the two processes would signal a problem.

Naming conventions were carefully developed and rigidly enforced in WoT. For example, weapon expenditures are tracked by weapon type using the variables "SumWpnsXpnd-Type" and "Sum2WpnsXpndType" to represent the sum of weapons expended and the sum of squares of the number of weapons expended.

WoT has thorough internal documentation describing and explaining code and even indicating possible code modifications. To illustrate, the method for planning strikes has over 200 lines, equating to over five pages, of internal documentation

As indicated above, WoT was tested across a variety of Windows, Unix, Linux, and Macintosh computers. It was tested using two C++ compilers for Windows and two C++ compilers for Macintosh computers. No code modifications were found to be needed to port the code from one machine to another, and results were identical across all machines.

Finally, WoT was strenuously tested across thousands of machine-generated cases with no inexplicable results reported.

Experimental Design

As with many RDM exercises, this project employed an XLRM framework to help guide model development and data gathering. In addition, the RDM analysis's participatory scoping step—organizing discussions within the project team and with the project sponsor—relied heavily on this framework. XLRM proves useful because it helps organize relevant factors into the components of a decision-centric analysis.

The letters X, L, R, and M refer to four categories of factors important to an RDM analysis:

- **Exogenous uncertainties (X)** are factors outside decision makers' control, such as the future security environment, funding levels, and the adversary's capabilities that influence the ability of a munitions mix strategy to achieve military goals.
- **Policy levers (L)** are near-term actions that decision makers want to consider—in this case the initial munitions mix and how munitions stockpiles will be replenished over time.
- **Relationships (R)**, generally represented by simulation models, describe how the policy levers perform, as measured by the metrics, under the various uncertainties.
- **Metrics (M)** are the performance standards used to evaluate whether or not a choice of policy levers achieves decision makers' goals.

In essence, RDM compares the performance of alternative combinations of policy levers, as evaluated by the metrics, over a wide range of uncertain futures using the relationships or models.

This appendix is organized around this XLRM framework, as summarized in Table 2.1. It first describes the simulation models, the relationships (R), used in this study. It then describes the specific metrics (M) used to judge the effectiveness of alternative munitions mix strategies, the policy levers (L) that constitute the specific munitions mix strategies considered in this study, and the exogenous uncertain factors (X) that might affect the performance of these strategies.

Relationships (R): Models

This project uses a WoT model, which is a fast-running emulator of the more detailed campaign models currently used by CAPE. This model is described in detail in Appendix A. As shown in Figure 2.4, we combine this model with a CG, in Analytica, that creates a series of campaigns for the WoT model. We summarize the models here and describe them in detail below.

The WoT model is the workhorse of the simulation process. It simulates activities in each campaign, day-to-day. It orders targets by their priority and each day matches the most effective weapons for the highest-priority targets, within a number of constraints.[1] The WoT model determines whether a target is destroyed by drawing a pseudorandom number between 0 and 1 and comparing it to the Pk for the combination of weapon and target chosen.[2] WoT runs the simulation multiple times (100 times in our simulations) and reports a number of results (e.g., the percentage of the simulated campaigns that were successful, the 95th percentile day the campaign was completed, and the 95th percentile number of each weapon expended). The CG, which is modeled in Analytica, is represented by the diagram in Figure B.1. The purpose of the CG is to model the dynamic aspects of the munitions planning process. For each two-year period, the CG reads the security environment and conflict characters (uncertain inputs provided to the CG for each period of the simulation) and identifies which campaigns will occur in that period. It then constructs the input files for each campaign.

Construction of most of the WoT inputs is mechanistic. For example, appropriate targets sets and Pk values are selected based on each type of campaign and may be modified according to the uncertainties provided as inputs. The key to the CG is the generation of weapons inventories. "Munitions Acquisition" acquires weapons based on available funding ("Defense Funding Level" is set based on the severity of the conflict character in the previous period) and the "Policy Levers" (described below) that dictate the rules for acquiring weapons. In addition to gaining weapons in inventories ("Global Munitions"), the CG accounts for weapons expended in previous campaigns. "Tactical Outcomes" serves as a function that constructs the final WoT

Figure B.1
Schematic of Campaign Generator in Analytica

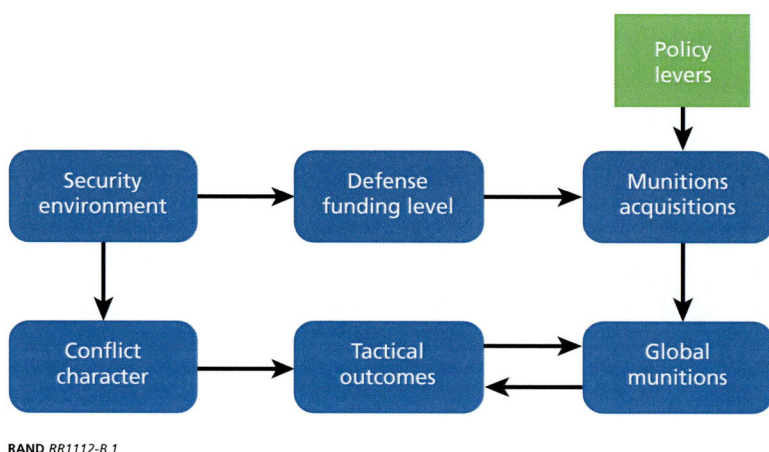

RAND RR1112-B.1

---

[1]   For example, the campaign may be nonpermissive until a certain group of targets is destroyed. In a nonpermissive environment, only a subset of weapons can be used.

[2]   Some targets consist of multiple DMPIs. For these targets, every DMPI needs to be destroyed before the entire target is destroyed.

input data, runs the WoT model, and generates an output from the WoT model that includes the updated munitions inventories.

To interface the CG with WoT, we wrote parser functions (in C++ and Java) that transform the WoT inputs from Analytica's output format into the proper format for WoT.[3] The parser functions execute the WoT model and transform WoT's outputs. Metrics (M) concerning the cost and success of that period's campaigns are stored in an Excel file, and outputs describing the remaining inventories are passed back to the CG so that it can determine inventories for future periods.

Measures (M)

We consider three measures for the performance of the munitions mix strategies:

1. the cost of munitions purchases over a 20-year time period
2. the total time needed to complete all campaigns
3. the percentage of campaigns that are successfully completed.

For each of these measures, we also considered regret calculations. For the total time needed to complete all campaigns and the percentage of campaigns that are successfully completed, regret for a specific case is the difference between a particular strategy's value of the metric and the best value of the metric across all strategies. For the cost of munitions, regret for a case is the difference between the cost of a particular strategy and the lowest cost of a strategy that has at least the same level of success (measured in metric 3).

Metric 3, the success rate, is the most important metric overall because it measures the ultimate goal of munitions purchases—to be successful in contingencies. Metrics 1 and 2 (cost and time) are both inputs that contribute to success. Decision makers would prefer to minimize both, but greater cost and greater time both contribute to greater success.

In the second round of runs, we also focused on a metric called "Total Cost" that combined cost and the time. We estimated that the cost of an additional day to complete a campaign was about 20 times the average cost of daily munitions expenditures.

Policy Levers (L)—First Round

The first round of the study considers three alternative munitions acquisitions strategies that vary across two dimensions.

We consider two sets of rules by which munitions acquisitions build inventories over time:

1. Stockpile (**Stock**)—the government spends all available funds on purchasing new munitions. Purchases are in proportion to portfolio goals.
2. Replenishment (**Rpln**)—the government uses available funding to return munitions stockpiles to their "optimal" levels. If insufficient funds are available for total replen-

---

[3] In addition, the parser updates some parameters based on uncertainties. For example, Analytica was very slow to update Pk values based on uncertainties, so the update was moved to the parser so that simulations could be conducted at an acceptable speed.

ishment, the government spends funds in proportion to the shortages from portfolio goals.[4]

We consider three portfolio goals toward which these rules acquire. Two of these portfolio goals have both time and cost-minimized variants, resulting in five total portfolio goals.

a)   Original portfolio (**Base**)—the goal is to purchase the same portfolio mix that initially existed.
b)   Big Wars (**Big**)—the goal is to purchase a portfolio mix optimized for two consecutive major regional campaigns:
      a)   Time-minimized
      b)   Cost-minimized
c)   Small War (**Small**)—the goal is to purchase a portfolio mix optimized for one major regional and COIN campaign
      a)   Time-minimized
      b)   Cost-minimized.

The combination of these two rules and three goals results in ten strategies:

1.   Stockpile—original portfolio (*Base-Stock*): Spend all available funds on new weapons to build stockpile with same weapons *proportions* as in the original portfolio.
2.   Stockpile—Big Wars—time-optimized (*Big-Time-Stock*): Spend all available funds on new weapons to build stockpile with same weapons *proportions* as in the time-optimized Big Wars portfolio.
3.   Stockpile—Big Wars—cost-optimized (*Big-Cost-Stock*): Spend all available funds on new weapons to build stockpile with same weapons *proportions* as in the cost-optimized Big Wars portfolio.
4.   Stockpile—Small War—time-optimized (*Small-Time-Stock*): Spend all available funds on new weapons to build stockpile with same weapons *proportions* as in the time-optimized Small War portfolio.
5.   Stockpile—Small War—cost-optimized (*Small-Cost-Stock*): Spend all available funds on new weapons to build stockpile with same weapons *proportions* as in the cost-optimized Small War portfolio.
6.   Replenishment—original portfolio (*Base-Rpln*): Purchase new weapons to return inventories to original numbers.
7.   Replenishment—Big Wars—time-optimized (*Big-Time-Rpln*): Purchase new weapons to bring inventories at least as high as in the time-optimized Big Wars portfolio.
8.   Replenishment—Big Wars—cost-optimized (*Big-Cost-Rpln*): Purchase new weapons to bring inventories at least as high as in the cost-optimized Big Wars portfolio.
9.   Replenishment—Small War—time-optimized (*Small-Time-Rpln*): Purchase new weapons to bring inventories at least as high as in the time-optimized Small War portfolio.
10. Replenishment—Small War—cost-optimized (*Small-Cost-Rpln*): Purchase new weapons to bring inventories at least as high as in the time-optimized Small War portfolio.

---

[4]   Initial inventories of weapons may exceed the optimal portfolio goals. When this happens, the replenishment strategy will keep those excess weapons but not purchase any additional.

To determine the time-optimal munitions mixes for the replenishment strategies, the WoT model includes a mode that attempts to minimize the time of campaign using a "greedy" algorithm. This algorithm is unconstrained by weapons inventories. Instead, in each period, it matches targets to the most effective weapons, regardless of cost or inventories. When the campaign is complete, it reports the average weapons used, which is the "optimal" portfolio to minimize time.

We also developed a mode in WoT that attempts to minimize weapons cost. It does this using a greedy algorithm that selects the most cost-effective weapons (highest Pk per dollar spent). The cost-minimization mode splits each campaign into two portions—the nonpermissive portion, in which only standoff weapons can be used, and the permissive portion, in which any weapon can be used. In the nonpermissive portion, WoT selects only targets that must be destroyed before the campaign becomes permissive, and WoT chooses cost-effective standoff weapons. In the permissive portion, WoT targets the remaining targets and mostly chooses cost-effective nonstandoff weapons, since they are much more cost-effective than standoff weapons.

The optimization process includes a number of assumptions that affect optimal inventories:

1. The optimization uses best guesses for weapons and targets data (e.g., Pk values). These best guesses do not account for uncertainties (see below) that change these values during the RDM process.
2. However, the optimization assumes that 100 percent of targets need to be destroyed, whereas the best guess is 80 percent (and a range of 60 to 100 percent).
3. The optimization process uses 1,000 trials, whereas campaigns in the RDM process use only 100 trials.
4. The optimization process assumes that WoT uses one weapon per target.[5]
5. The optimization allows any completion time, even if it is above the 200 days allowed in our later analysis.

We specify two sets of expected campaigns for determining optimal weapons portfolio requirements. The campaigns used to define munitions mixes are the following:

1. Big War, with two nearly simultaneous, medium-size conflicts (i.e., two major regional campaigns).
2. Small War, with prolonged irregular campaigns punctuated by numerous precision strikes (i.e., one major regional and COIN campaign).

We initially considered choosing both a risk averse portfolio level (where we have 90 percent confidence that the portfolio would be successful against our expected campaign(s)) and a risk tolerant portfolio level (where we have a 50 percent confidence that the portfolio would be successful). However, during initial development of the strategies, we found that the differences in the inventories for risk averse and risk tolerant was very small—usually less than a 1 percent difference and always less than 2 percent. We decided that given such small differences, it was unlikely that decision makers would accept any risk. Instead, we decided to use a single,

---

[5] Other modes have been developed in WoT that can use multiple weapons per target, but they have not been implemented in this analysis.

risk averse portfolio with a 5 percent buffer above the optimal portfolio level. This 5 percent buffer ensures that the portfolio will always be successful against the expected campaign(s).[6]

Table B.1 shows the five weapons goals used in the first round of the analysis. *Big-Time* uses time minimization to plan for Big Wars. *Small-Time* uses time minimization to plan for a Small War. *Big-Cost* uses cost minimization to plan for Big Wars. *Small-Cost* uses cost minimization to plan for a Small War. Finally, *Baseline* gives our initial estimates of current inventories. Each optimized goal contains nonzero inventories for only a handful of weapons. Because they were developed with a goal of minimizing the time of campaigns, time-minimized inventories tend to be highly effective weapons that can be used in nonpermissive environments. Because time optimization ignores cost considerations, time-minimized inventories also tend to be expensive (e.g., both *Big-Time* and *Small-Time* require many massive ordnance penetrators—MOPs). *Big-Cost* also purchases standoff weapons because these are necessary in the nonpermissive phase of the Big Wars. However, it buys many fewer than even the *Small-Time* portfolio goals. Although the time-minimized inventories bought few, if any, nonstandoff weapons, the cost-minimized inventories are heavy in nonstandoff weapons since they are so inexpensive.

Policy Levers (L)—Second Round

In successive rounds of RDM analysis, strategies are reevaluated to try to improve on the performance of strategies from previous rounds. After seeing the results of the first round and discussing the results within the project team and with the sponsor, we decided on several changes in strategies for the second round:

- **Elimination of stockpile strategies**. Stockpile strategies rarely resulted in substantial improvements in performance over replenishment strategies, but usually cost a lot more. In addition, stockpile strategies were seen as unrealistic, since it would be politically difficult to buy more than the portfolio goals.
- **Addition of "previous expenditure" strategies**. This strategy purchases weapons in proportion to how many were expended during campaigns in the previous period. *The previous expenditure strategy was implemented in the CG but not in time to analyze it and validate it during the project. Initial results appear similar to the replenishment strategy.*
- **Elimination of Small War goals**. Small War strategies were very unsuccessful in the first round. Their inventories were too small to be successful against larger wars, and the Small War was always permissive, so weapons inventories geared to the Small War had inadequate standoff weapons.
- **Consideration of deterrence campaigns in portfolio optimization**. A weakness of the first-round strategies was in deterrence campaigns. Although these campaigns are relatively small, they rely on high-end weapons.
- **Consideration of total cost.** Previous portfolio goals attempted to minimize either cost or time. New goals attempt to minimize total cost—a combination of cost and time minimization.

---

[6] The optimal inventories are based on the average weapons used; therefore, about half the campaigns would run out of weapons without a buffer.

Table B.1
Portfolio Goals Used in First Phase Strategies

ID	Weapon	Big-Time	Small-Time	Big-Cost	Small-Cost	Baseline
W00	AGM-130C (BLU 109)	7,088	0	0	0	250
W01	AGM-130A (Mk 84)	40,709	564	886	0	250
W02	AGM-154 (JSOW)	27,430	8,174	2,104	0	24,000
W03	AGM-158 (JASSM-ER)	3,944	0	0	0	2,400
W04	AGM-64 (Maverick)	2	0	6,292	6,301	40,000
W05	AGM-84H (SLAM/ER)	0	0	0	0	6,000
W06	AGM-86 (CALCM)	0	0	0	0	1,100
W07	AGM-88E (HARM/AARGM)	0	0	935	0	19,000
W08	BGM-109E (TLAM)	2,787	502	0	0	2,000
W09	GBU-10 Paveway II (BLU 109)	0	0	2,144	0	5,500
W10	GBU-10 Paveway II (Mk 84)	0	0	13,123	0	5,500
W11	GBU-12 (Paveway II)	0	0	17,850	3,152	32,000
W12	GBU-15 (BLU-109)	0	200	0	0	1,400
W13	GBU-15 (Mk 84)	0	0	5,019	664	1,400
W14	GBU-24 Paveway III (BLU-109)	0	0	0	0	4,300
W15	GBU-24 Paveway III (BLU-116)	0	0	2,098	1,889	4,300
W16	GBU-24 Paveway III (Mk 84)	0	0	2,627	0	4,300
W17	GBU-28 Bunker Buster	0	0	0	0	300
W18	GBU-31 JDAM (2,000 lb)	0	158	0	0	29,200
W19	GBU-32 JDAM (1,000 lb)	0	0	0	0	29,200
W20	GBU-36/37 (GAM)	21,758	0	42,011	0	10
W21	GBU-38 JDAM (500 lb)	0	441	0	0	29,200
W22	GBU-39 SDB	0	0	0	2,522	24,000
W23	GBU-53B SDB II	0	0	0	0	17,000
W24	GBU-54 (Laser JDAM)	0	0	8,528	1,846	240,000
W25	GBU-57B MOP	1,905	351	0	0	20
W26	MGM-168 ATACMS	3,037	345	0	0	2,000
W27	Mk-82	0	0	0	0	500,000
W28	Mk-83	0	0	12,604	3,154	500,000
W29	Mk-84	510	0	9,458	0	500,000
	Total cost ($ billions)	$66.7	$8.7	$14.4	$1.3	$73.1

NOTE: Shading indicates that the weapon is a standoff (PGM) weapon.

- **Addition of goals that are robust to GPS degradation.** The first round of analysis revealed that GPS degradation was a key determinant of success.

In the second round of analysis, we consider two sets of rules by which munitions acquisitions build inventories over time:

1. Replenishment (***Rpln***)—the government uses available funding to return munitions stockpiles to their "optimal" levels. If insufficient funds are available for total replenishment, the government spends funds in proportion to the shortages from portfolio goals.
2. Previous expenditure (***Prev***)—the government uses available funding to purchase weapons in proportion to their expenditure in the previous period. Munitions are purchased up until the stockpiles are returned to "optimal" portfolio goals.

If funding is sufficient to purchase enough munitions to return to the optimal portfolio goals, both replenishment and previous expenditure are equivalent—they fully restore the stockpile to its desired level. However, if funding is not sufficient to return to the portfolio goals, the two rules will purchase different shares of weapons.

We consider six portfolio goals, which represent the ideal stockpiles toward which these rules build. Three of these portfolio goals are carried over from the first round, and three are attempts to minimize total costs through linear combinations of time-minimized and cost-minimized inventories.[7]

a) Original portfolio (*Baseline*)—the goal is to purchase the same portfolio mix that initially existed (*same as first round*)
b) Big Wars (*Big*)—the goal is to purchase an portfolio mix optimized for two consecutive major regional campaigns (*same as first round*)
    a) Time-minimized
    b) Cost-minimized
c) Total cost-minimization, Big Wars, and deterrence (*Big+Deter-Time*)—a combination designed to minimize total costs (i.e., both munitions costs and the cost of days to complete a campaign) that is 95 percent time-minimization inventories and 5 percent cost-minimization inventories. Both inventories account for two consecutive major regional campaigns plus a deterrence campaign.
d) Total cost-minimization, Big Wars, and deterrence (*Big+Deter-Mixed*)—a combination designed to minimize total that is 55 percent time-minimization inventories and 45 percent cost-minimization inventories. Cost-minimization inventories are heavy in stand-off weapons. Both inventories account for two consecutive major regional campaigns plus a deterrence campaign.
e) Total cost-minimization, Big Wars, and deterrence with degraded GPS (*Big+Deter/ GPS-Mixed*)—a combination designed to minimize total that is 55 percent time-minimization inventories and 45 percent cost-minimization inventories. Cost-minimization inventories are heavy in standoff weapons. Both inventories account for

---

[7]   Linear combinations mean that new portfolio goals are created by multiplying other portfolio goals and combining them. This method of minimizing total costs can be seen as a "third-best" solution to the total cost-optimization problem. The best solution would be to have WoT minimize internal costs internally, but such an improvement in WoT is beyond the resources of the project. The second-best solution would be for WoT to use heuristics, as it does with cost and time minimization. For example, instead of selecting the weapon with the best Pk, WoT could penalize weapons for their cost and the additional days to complete a campaign they may cause. This solution also required too many resources and created too much risk for this stage of the project. Therefore, we used linear combinations. Other types of linear combinations (e.g., combining inventories on a weapon-by-weapon basis so that one weapon could have a 0.05 factor, another a 0.25 factor, etc.) were not considered but could result in reductions in total cost.

two consecutive major regional campaigns plus a deterrence campaign, both of which have a 33 percent degradation in the performance of GPS weapons.

The combination of these two rules and six goals results in ten strategies:

1. Replenishment—original portfolio (*Baseline-Rpln*): Purchase new weapons to return inventories to original numbers (*same as first round*).
2. Replenishment—Big Wars—time-optimized (*Big-Time-Rpln*): Purchase new weapons to bring inventories at least as high as in the time-optimized Big Wars portfolio (*same as first round*).
3. Replenishment—Big Wars—cost-optimized (*Big-Cost-Rpln*): Purchase new weapons to bring inventories at least as high as in the cost-optimized Big Wars portfolio (*same as first round*).
4. Replenishment—Big Wars/deterrence—total cost-optimized (*Big+Deter/Time-Rpln*): Purchase new weapons to bring inventories at least as high as in the Big Wars/ deterrence, total cost-minimized portfolio.
5. Replenishment—Big Wars/deterrence—total cost-optimized (*Big+Deter/Mixed-Rpln*): Purchase new weapons to bring inventories at least as high as in the Big Wars/ deterrence, total cost-minimized portfolio, with heavy standoff weapons.
6. Replenishment—Big Wars/deterrence/GPS—total cost-optimized (*Big+Deter/GPS-Mixed-Rpln*): Purchase new weapons to bring inventories at least as high as in the Big Wars/deterrence with GPS degradation, total cost-minimized portfolio.
7. Previous expenditure—original portfolio (*Base-Prev*): Purchase new weapons in proportion to weapons use in the previous period until inventories returned to original numbers.
8. Previous expenditure—Big Wars—time-optimized (*Big-Time-Prev*): Purchase new weapons in proportion to weapons use in the previous period until inventories are at least as high as in the time-optimized Big Wars portfolio.
9. Previous expenditure—Big Wars—cost-optimized (*Big-Cost-Prev*): Purchase new weapons in proportion to weapons use in the previous period until inventories are at least as high as in the cost-optimized Big Wars portfolio.
10. Previous expenditure—Big Wars/deterrence—total cost-optimized (*Big+Deter/Time-Prev*): Purchase new weapons in proportion to weapons use in the previous period until inventories are at least as high as in the Big Wars/deterrence, total cost-minimized portfolio.
11. Previous expenditure—Big Wars/deterrence—total cost-optimized (*Big+Deter/Mixed-Prev*): Purchase new weapons in proportion to weapons use in the previous period until inventories are at least as high as in the Big Wars/deterrence, total cost-minimized portfolio, with heavy standoff weapons.
12. Previous expenditure—Big Wars/deterrence/GPS—total cost-optimized (*Big+Deter/GPS-Prev*): Purchase new weapons in proportion to weapons use in the previous period until inventories are at least as high as in the Big Wars/deterrence with GPS degradation, total cost-minimized portfolio.

The first three portfolio goals (a, b.a, and b.b) are carryovers from the first round. Those portfolio goals were listed in Table B.1.

New portfolio goals were created in WoT to account for the addition of the deterrence campaign. Developing these goals followed the same procedure used in the first round. As in the first round, those goals have a 5 percent buffer. These goals are listed in columns i and ii of Table B.2.

Table B.2
New Time- or Cost-Minimized Goals Used to Construct Second Phase Strategies (A)

ID	Weapon	(i) *Big/ Deter-Time*	(ii) *Big/ Deter-Cost*	(iii) *Big/ Deter-Standoff Cost*	(iv) *Big/Deter/ GPS-Time*	(v) *Big/Deter/ GPS-Cost*
W00	AGM-130C (BLU 109)	7,168	36	2,556	10,087	53
W01	AGM-130A (Mk 84)	41,327	986	7,399	60,527	1,479
W02	AGM-154 (JSOW)	28,809	3,283	25,314	42,671	4,931
W03	AGM-158 (JASSM-ER)	4,140	0	0	6,817	0
W04	AGM-64 (Maverick)	3	6,450	6,450	3	6,459
W05	AGM-84H (SLAM/ER)	0	0	0	0	0
W06	AGM-86 (CALCM)	0	0	0	0	0
W07	AGM-88E (HARM/AARGM)	0	987	987	0	1,481
W08	BGM-109E (TLAM)	3,053	0	0	4,860	0
W09	GBU-10 Paveway II (BLU 109)	0	2,144	2,144	0	2,144
W10	GBU-10 Paveway II (Mk 84)	0	13,460	13,460	0	15,077
W11	GBU-12 (Paveway II)	0	17,974	17,974	0	17,981
W12	GBU-15 (BLU-109)	0	0	0	0	2,262
W13	GBU-15 (Mk 84)	0	5,151	5,151	0	5,151
W14	GBU-24 Paveway III (BLU-109)	0	0	0	0	0
W15	GBU-24 Paveway III (BLU-116)	0	2,098	2,098	0	2,098
W16	GBU-24 Paveway III (Mk 84)	1	2,726	2,726	1	2,724
W17	GBU-28 Bunker Buster	0	0	0	0	0
W18	GBU-31 JDAM (2,000 lb)	0	0	0	0	0
W19	GBU-32 JDAM (1,000 lb)	0	0	0	0	0
W20	GBU-36/37 (GAM)	22,253	43,146	71,512	32,863	64,770
W21	GBU-38 JDAM (500 lb)	0	0	0	0	0
W22	GBU-39 SDB	0	0	0	0	0
W23	GBU-53B SDB II	0	0	0	0	0
W24	GBU-54 (Laser JDAM)	0	8,630	8,630	0	6,298
W25	GBU-57B MOP	2,073	88	1,489	3,185	132
W26	MGM-168 ATACMS	3,437	0	1,577	5,217	0
W27	Mk-82	0	0	0	0	0
W28	Mk-83	0	12,604	12,604	0	12,592
W29	Mk-84	528	9,873	9,873	557	9,864

NOTES: Shading indicates that the weapon is a standoff (PGM) weapon. These goals are used in linear combinations to construct goals actually used in the second phase.

To generate the portfolio goals when cost-minimization is standoff-weapon heavy (in portfolio goal d), we developed an additional set of portfolio goals using the optimization procedure from the previous period. The only change was that the campaigns were designed so that they never became nonpermissive. In other words, only standoff weapons would be effective. Then, for each weapon, we chose the maximum of the previous cost-minimization and the nonpermissive cost-minimization. This procedure assured that the standoff heavy cost minimization has sufficient quantities of cost-effective standoff weapons, but it also has large quantities of cheap, nonstandoff weapons as backup. We decided to use a standoff-heavy cost-minimization portfolio goal because in the first round we found that WoT would often use up most standoff weapons in the first campaign of consecutive campaigns, leaving few standoff weapons for later campaigns, thus dooming those campaigns to failure.[8] This goal is listed in column iii of Table B.2.

To develop portfolio goals that account for GPS degradation, the same procedure was used, with one significant change. Rather than using the regular Pk table (see the sections below for more discussion on the Pk table and adjustments to it), we used a Pk table that was adjusted so that all GPS weapons were 33 percent less effective. These goals are listed in columns iv and v of Table B.2.

All of the new portfolio goals (c, d, and e) used in the second phase are linear combinations of the time- and cost-minimized inventories discussed in the previous paragraphs and listed in Table B.2, according to the following linear combinations that attempt to minimize total costs:

Big+Deter-Time
Big+Deter-Mixed
Big+Deter/GPS-Mixed.

These goals, which are the goals that are actually used in the second phase of the analysis, are listed in Table B.3.

To discover the linear combination of portfolio goals that minimized total costs (i.e., munitions cost and the cost of days to completion), we developed a new procedure that ran in WoT a double major regional campaign[9] using a weapons portfolio that was a linear combination with percentage shares (e.g., 95 percent and 5 percent) that are divisible by 5. We plotted the resulting cost and days to completion on a chart and compared those to an isocost line that valued each day at about $7 billion (roughly 20 times the average daily cost of munitions). Each campaign in WoT ran for only 100 trials to conserve time.[10]

---

[8]   This behavior is a result of WoT's underlying behavior; it always picks the most effective weapon, regardless of cost. Further, WoT does not consider the opportunity cost of using weapons now instead of saving them for another campaign. With cost-minimized inventories, the opportunity cost of using standoff weapons can be huge because failing to save them means that later campaigns might run out of standoff weapons and fail before a campaign becomes permissive.

[9]   Instead of two consecutive major regional campaigns, we doubled the number of targets in one major regional campaign and allowed 400 days for completion. We did this because two consecutive major regional campaigns tend to disfavor cost-minimizing goals because WoT expends standoff weapons in the first campaign that need to be saved for the second campaign. This decision tends to increase slightly the share of cost-minimization inventories in the total-cost-minimized weapons goals.

[10]   More fidelity would be possible with more trials and using linear combinations that are not divisible by 5 percent, e.g., 96/4.

Table B.3
New Time- or Cost-Minimized Goals Used to Construct Second Phase Strategies (B)

ID	Weapon	(C) Big+Deter-Time	(D) Big+Deter-Mixed	(E) Big+Deter/GPS-Mixed
W00	AGM-130C (BLU 109)	6,811	5,093	7,579
W01	AGM-130A (Mk 84)	39,310	26,059	45,765
W02	AGM-154 (JSOW)	27,533	27,236	33,236
W03	AGM-158 (JASSM-ER)	3,933	2,277	5,113
W04	AGM-64 (Maverick)	325	2,904	1,617
W05	AGM-84H (SLAM/ER)	0	0	0
W06	AGM-86 (CALCM)	0	0	0
W07	AGM-88E (HARM/AARGM)	49	444	370
W08	BGM-109E (TLAM)	2,900	1,679	3,645
W09	GBU-10 Paveway II (BLU 109)	107	965	536
W10	GBU-10 Paveway II (Mk 84)	673	6,057	3,769
W11	GBU-12 (Paveway II)	899	8,088	4,495
W12	GBU-15 (BLU-109)	0	0	566
W13	GBU-15 (Mk 84)	258	2,318	1,288
W14	GBU-24 Paveway III (BLU-109)	0	0	0
W15	GBU-24 Paveway III (BLU-116)	105	944	525
W16	GBU-24 Paveway III (Mk 84)	137	1,227	682
W17	GBU-28 Bunker Buster	0	0	0
W18	GBU-31 JDAM (2,000 lb)	0	0	0
W19	GBU-32 JDAM (1,000 lb)	0	0	0
W20	GBU-36/37 (GAM)	23,298	44,420	40,840
W21	GBU-38 JDAM (500 lb)	0	0	0
W22	GBU-39 SDB	0	0	0
W23	GBU-53B SDB II	0	0	0
W24	GBU-54 (Laser JDAM)	432	3,884	1,575
W25	GBU-57B MOP	1,974	1,810	2,422
W26	MGM-168 ATACMS	3,265	2,600	3,913
W27	Mk-82	0	0	0
W28	Mk-83	630	5,672	3,148
W29	Mk-84	995	4,733	2,884
	Total cost ($ billions)	67.2	60.6	84.5

NOTE: Shading indicates that the weapon is a standoff (PGM) weapon.

Figure B.2 shows the munitions cost and time tradeoff when combining portfolio goals (i) and (ii). The blue dots indicate that the weapons portfolio is successful in 100 percent of trials. Red dots indicate that the portfolio is successful in less than 100 percent of trials. Inventories that are heavy on the cost-minimizing inventories (ii) are never successful, so they are not included on the chart. The blue lines are isocost lines that value each day of a campaign

Figure B.2
***Big+Deter-Time* Total Cost Minimization: Cost/Time Tradeoff for Linear Combinations of (i) and (ii)**

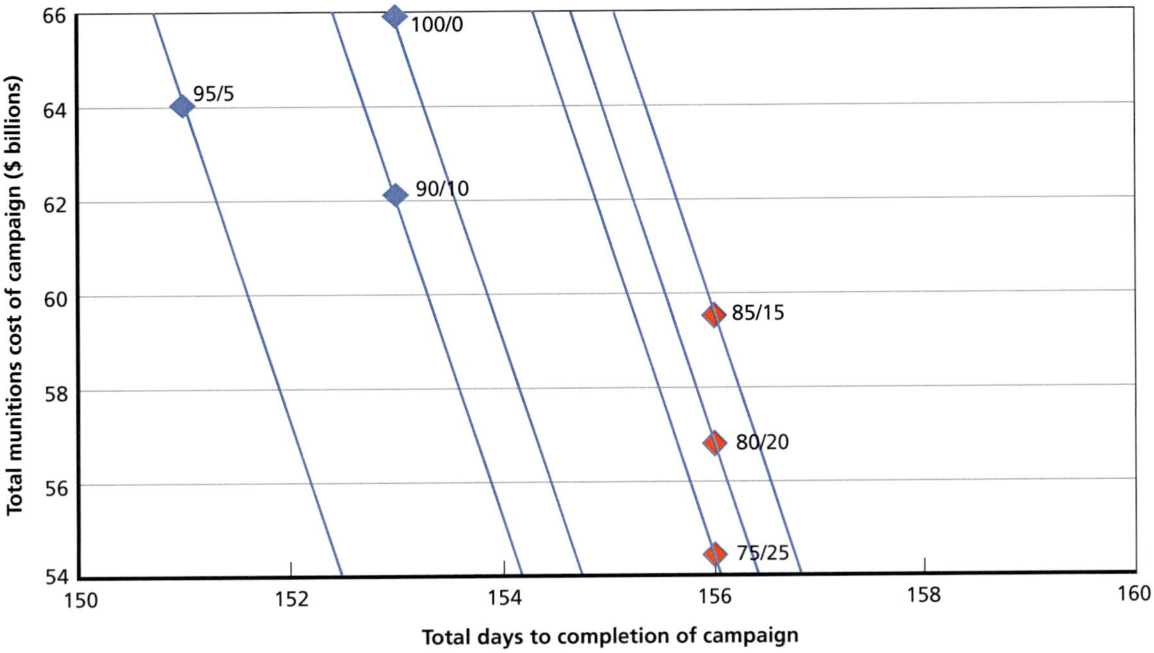

at about $7 billion. It is clear that the "95/5" goal minimizes total cost (i.e., its isocost line is nearest the origin). This point represents a 95 percent mix of (i) and a 5 percent mix of (ii).[11]

Figure B.3 shows a similar chart for combinations of (i) and (iii). Note that when the cost-minimizing goals are heavier in standoff weapons (iii), combinations with a greater share of the cost-minimizing inventories tend to do better. The lowest total cost appears in the 55/45 combination. Comparing this to the 95/5 combination in Figure B.2, we see that the 55/45 combination is cheaper ($54 billion vs. $64 billion) and takes less time to complete campaigns (147 days vs. 151 days). This suggests that portfolio goals (D) should dominate portfolio goals (C), but the ultimate arbiter of strategy effectiveness is when the cases are run across many years with many different uncertainties.

Finally, Figure B.4 shows combinations of (iv) and (v). WoT simulations include the 33 percent degradation in Pk values of GPS weapons, which leads to substantially increased costs of munitions (more munitions have to be used to destroy the same number of targets) and increases the number of days to completion (more misses means that it takes longer to destroy all targets). The isocost lines in this chart show that the 75/25 combination has the lowest total cost.

---

[11] These results clearly favor the time-minimizing portfolio (i). However, the 100/0 mixture does not do as well as the 95/5 mixture. The problem is that the time-minimization portfolio has trouble with one type of target (depots). For this target, the Pk table says that a dumb bomb, the Mk-84, is the best weapon, but time-minimizing inventories include very few non-standoff weapons. The Pk of the Mk-84 for this target is only 0.20, so there is high variability in how many weapons it takes to destroy all the targets; therefore, the cost-minimizing portfolio often runs out of Mk-84s and has to use a second-best weapon. A 95/5 combination doubles the number of Mk-84s, which prevents Mk-84s from running out.

Figure B.3
Big+Deter-Mixed **Total Cost Minimization: Cost/Time Tradeoff for Linear Combinations of (i) and (iii)**

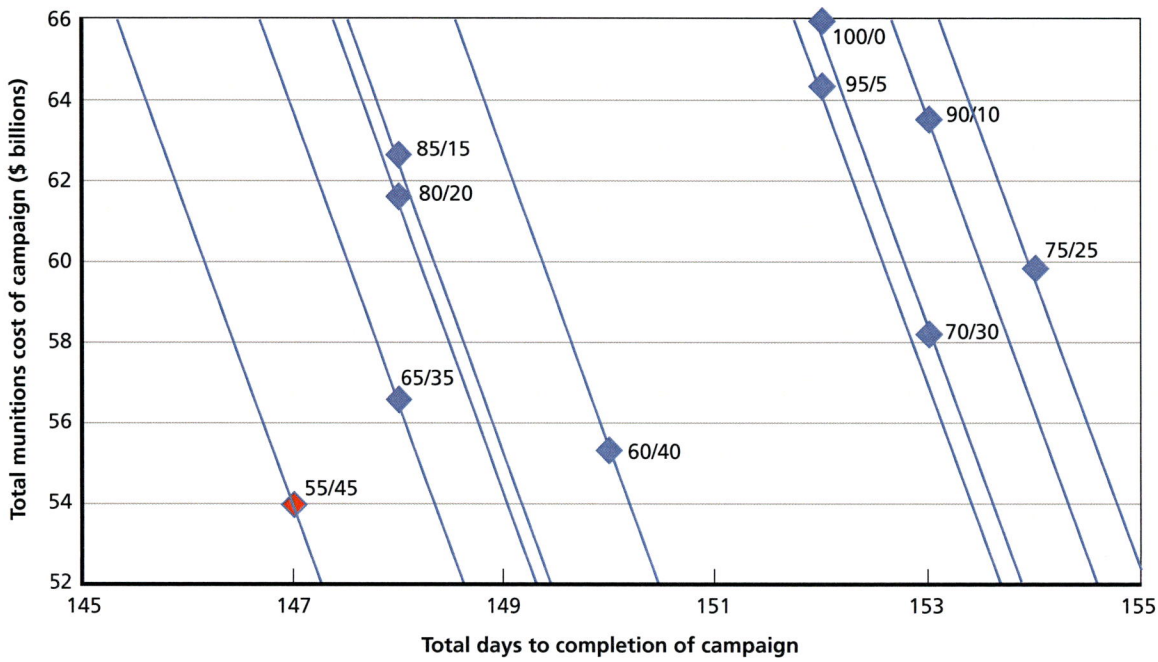

Figure B.4
Big+Deter/GPS-Mixed **Total Cost Minimization: Cost/Time Tradeoff for Linear Combinations of (iv) and (v)**

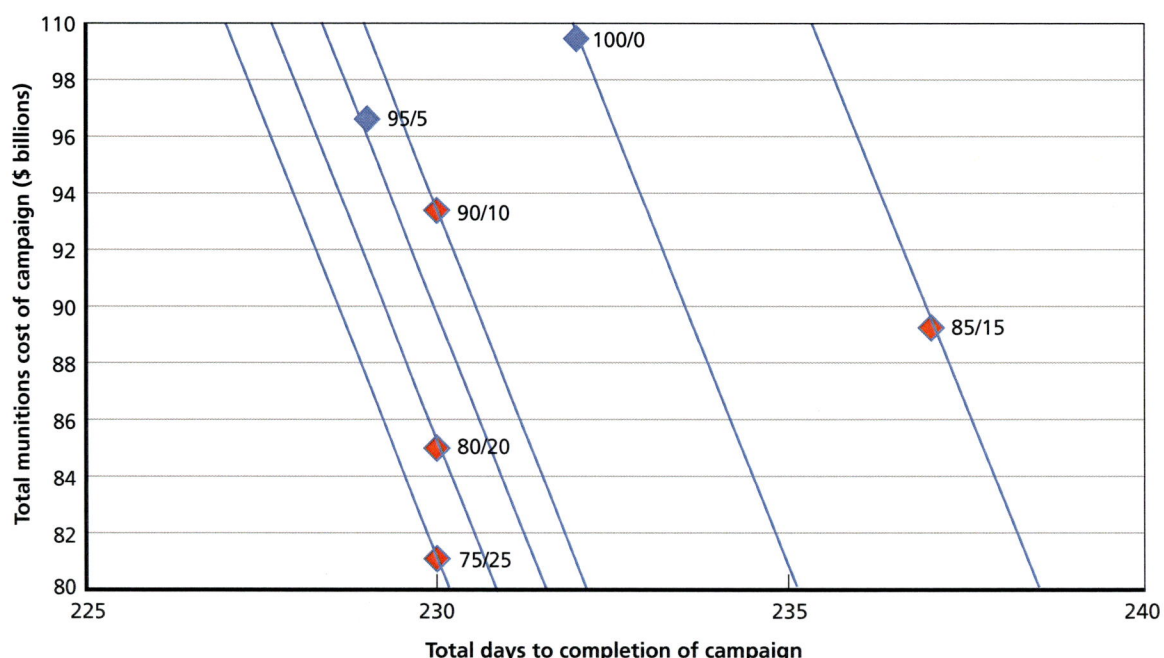

Exogenous Uncertainties (X)

This study aims to consider the performance of alternative munitions mix strategies over a wide range of future conditions. The study focused on several categories of uncertain factors, including the following.

Security Environment

Security environments show the sequence and type of security conditions over all periods. For the first phase, we used 20 security environments. Nine of them are notional security environments that were constructed early in the project. To obtain the other 11, we developed a spreadsheet that randomly creates security environments based on the rules of the different security conditions (e.g., global war lasts for at least two periods). Metrics were created that measured each security environment in three ways:

- **Average severity**: Each security condition is assigned an integer between 1 and 6 based on the relative severity of the conflict (e.g., global war is assigned a 6). For each security environment, the magnitude is calculated by adding values for each period, and the magnitude is normalized (e.g., if the security condition is always quiescent, its magnitude is 0, and if the security condition is always global war, its magnitude is 1). Magnitudes above 0.60 were excluded.
- **Variability**: Using the assigned integers, variability was calculated by the adding the absolute value of changes between each period. Variability was then normalized so it would fall between 0 and 1.
- **Trend**: Using the assigned integers, the trend (i.e., the slope of the linear regression of the integers over time) was calculated. The trend was normalized so it would fall between 0 and 1.

The objective of selecting the other 11 security environments was to maximize the dissimilarity in these three metrics. This dissimilarity was measured by adding the Euclidean distance between each possible pair of security environments.[12] The spreadsheet generated 100,000 sets of security environments, and the set with the highest Euclidean distance was chosen. These security environments are provided in Table B.3.

After analyzing the results of the first phase of the project, we realized that we did not have a security environment that closely approximated the past 20 years. SE 21 is our estimate of a security environment that best approximates the past 20 years.

Also, after analyzing the results of the first phase of the project, we realized that the additional 11 security environments that were added tended to be fairly severe, and very often none of our strategies were consistently successful. We added four additional security environments using a similar procedure as shown above but oversampled less-severe security conditions.[13] In addition, in the new security environments, we manually changed any security condition F

---

[12] Euclidean distance takes the square root of the sums of the squares of the difference between each of the three variables.

[13] The problem with the previous procedure was that we allowed all security conditions to be weighted equally. According to the law of large numbers, the average severity of 11 security environments with 10 security conditions each will tend to be close to 0.5, and large deviations will rarely be realized. The previous procedure excluded any security environments with an average severity of 0.6 or above to counter extreme security environments where success would be impossible, but this exclusion did not result in many less-severe security environments.

(global war) to security condition E (multiple major regional) because global wars proved so severe that given the funding limitations in the model, no strategy could ever be successful.

Figure B.5 shows the distributions of the three attributes used to create the security environments. The initial security environments clearly were concentrated in a moderate range. The additional security environments helped to fill gaps in lower severity.

Weapons Effectiveness

Pk values were modified to account for two uncertainties:

- **GPS-driven changes**: GPS weapons have their Pk varied ± 50 percent accounting for uncertainties about changes in GPS technology or GPS countermeasures.
- **Blinding terminal guidance weapons**: PGMs with seeker heads have a Pk of 0 when "zapped" by an adversary. Zapping frequency is varied between 0 and 75 percent of the time they are launched. To account for this blinding, the Pks are modified so that they account for both the Pk and the probability that the PGM is not zapped.

Uncertainty in Adversaries' Air Defense Capability

An additional uncertainty accounts for adversaries' air defense capabilities.

The models currently assume that most campaigns begin in a nonpermissive environment and do not become permissive until all targets above a certain priority level have been destroyed (i.e., when the adversary's SAMs have been destroyed). This priority level is unique

Figure B.5
Distributions of Attributes for the 25 Security Environments

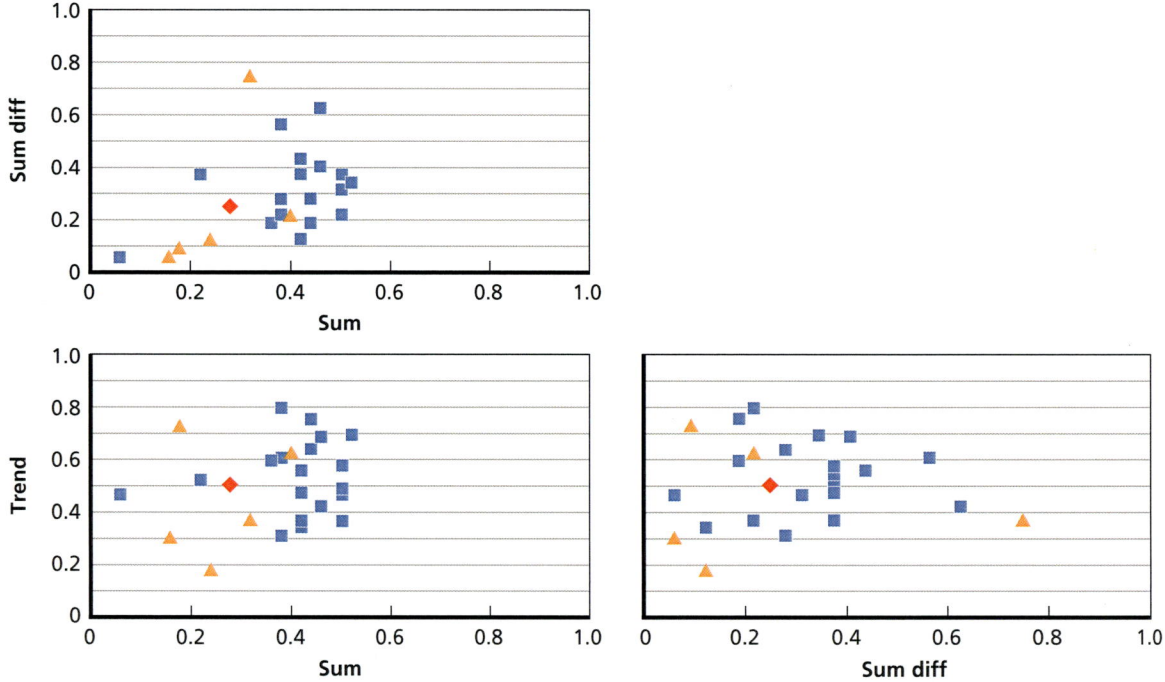

NOTE: Blue indicates the SE used in the first phase; red indicates SE 21, which approximates the past 20 years; orange indicates that additional SEs were added in the second phase.
RAND *RR1112-B.5*

to every type of campaign. To model the uncertainty in adversaries' air defense capabilities, we change the number of SAMs by ± 50 percent.

For example, a massive regional campaign has 29 types of targets, each with its own priority level (i.e., there are 29 priority levels). The SAMs are the 7th highest priority, so we assume that once the top seven priorities are destroyed completely, the campaign becomes permissive and all weapons may be used. If this uncertainty were + 50 percent, then 50 percent more SAMs would need to be destroyed before the campaign became permissive, which would probably use many more expensive, standoff weapons.

Uncertainty in Adversaries' Political Will

The WoT model allows users to specify a certain percentage of all targets that must be destroyed to compel an adversary to surrender. We assume that this percentage is the same across all campaigns, but we acknowledge that there could be substantial uncertainty. Thus, the range of values for adversaries' political will is 80 percent ± 20 percent; i.e., it can range from 60 percent to 100 percent.

Uncertainty in Weapons Delivery Rates

Weapons are currently assigned to one of four delivery modes (nonstealth, stealth, naval, and land). Each delivery mode is assumed to have surge delivery rate that lasts for a certain period of time and a lower, steady-state delivery rate that follows. Delivery rates are unique to each campaign and are a percentage of the maximum delivery rates.

Uncertainty in weapons delivery rates is accounted for by modifying the maximum delivery rates (both surge and steady-state) by a factor of ± 50%. This uncertainty accounts both for uncertainties in our modeling and estimates of weapons delivery and for uncertain attributes of campaigns that will affect delivery rates. For example, a campaign that requires delivery vehicles to be based long distances away will have a lower weapons delivery rate than a campaign where delivery vehicles are based short distances away.

Uncertainty in Munitions Funding (Not Implemented)

The munitions acquisition node in Analytica assumes that 0.85 percent of DoD funding can be allotted to munitions acquisition. We hope to incorporate an uncertainty factor that modifies that funding amount by ± 50 percent. This uncertainty factor would account for uncertainties in:

- overall levels of GDP for the United States
- the percentage of GDP allocated for DoD
- the percentage of DoD funding allocated for munitions acquisition
- the overall price level of munitions.

Because of complications in the Analytica design, we did not implement this uncertainty.

Uncertainty About the Fog of War (Not Implemented)

The WoT model currently assumes that the success of strikes can be observed instantaneously. An uncertainty that may be incorporated into a future version of WoT is uncertainty about whether a strike was successful. Such uncertainty could lead to retargeting of targets that have already been destroyed or waiting some amount of time to retarget targets where strikes have failed.

Model uncertainties are summarized in Table B.4.

Table B.4
Uncertainties Considered in the Analysis

Uncertainty	Procedure	Input	Comments
1. Security environment	Choose one of the predetermined SEs	SE is an integer between 1 and 20 (first phase) and between 1 and 25 (second phase)	
2. GPS technology	Adjust Pk of GPS weapons: If GPS < 0, Pk × (1 + GPS), else, Pk + (1 − Pk) × GPS	GPS is number, ± 0.50	
3. Blinding (B)	Adjust Pk of PGMs with seeker heads: Pk × (1 − B)	B is number, 0 to 0.75	(1 − B) is the chance weapon is not blinded
4. Adversary's air-defense capability (PERM)	Adjust the number of SAMs for each campaign: Original number of SAMs × (1 + PERM)	PERM is a percentage, ± 75%[a]	Will not affect campaigns without SAMs or those that begin as permissive
5. Adversary's political will (WILL)	Adjust the percentage of targets necessary to destroy to end the campaign: 80% + WILL	WILL is percentage, ± 20%	
6. Delivery rate (DR)	Adjust delivery rates (both surge and steady-state) by DR	DR is percentage, ± 50%	Accounts for uncertainties in such factors as geography
7. Munitions funding (MF)	*Adjust funding as a percentage of defense budget: 0.85% × MF*	*MF is percentage, ± 50%*	*Not implemented in Analytica*
8. Enhanced camouflage, concealment, and deception (FOG)	*TBD*	*TBD*	*Not implemented in WoT*

NOTE: Italics indicate model parameters not yet considered as uncertain in WoT.
[a] In the second phase, the range used was ± 75 percent.

Experimental Design

The first uncertainty is an unordered integer parameter with 20 (or 25) possible values. Uncertainties 2 to 6 are real numbers. We can thus conduct a Latin Hypercube design over the entire set of six parameters. Latin Hypercube is a commonly used method for generating a sample of futures to run in computational experiments. In a Monte Carlo design, futures are chosen randomly over the full range of uncertainties. Latin Hypercube also chooses futures randomly but in such a way that no two futures can have values for any uncertain parameter too close to one another. (The technical term is that Latin Hypercube uses a quasi-random design.) Thus, for any finite set of futures, a Latin Hypercube design ensures a more complete and more uniform sampling than a Monte Carlo design.

In the first phase of the analysis, we constructed a 1,000 future sample (20 security environments with 50 futures each).[14] We consider 10 weapons acquisition strategies. Therefore, we ran 10 strategies × 1,000 futures/strategy = 10,000 cases.

In the second phase of the analysis, we conducted a Latin Hypercube design over uncertainties 2 to 6, resulting in 50 futures. We repeated each set of 50 futures across all 25 security

---

[14]   In the first phase, we constructed a Latin Hypercube of 1,000 cases, which allowed the 50 cases associated with each security environment to be different. To improve statistical comparisons across security environments in the second phase, we constructed a Latin Hypercube of 50 cases that were repeated in each of the 25 security environments.

environments for a total of 1,250 futures per strategy. We considered 12 weapons acquisition strategies, i.e., 12 strategies × 1,250 futures/strategy = 15,000 cases.

The inputs (Xs and Ls) for each case are then run through the simulation model to calculate the corresponding values of measures. The resulting set of model inputs and outputs are then gathered together in a database of the form shown in Table B.5, where each entry represents one case. Such a database can then be used to support the analyses presented in Chapter Three.

Lessons Learned

In regard to methodological lessons learned during the research project, one particularity stands out. We might have been able to run more iterations if we had started the RDM process with an analytical model instead of the WoT model. Starting with the WoT model had two significant costs that slowed down our development and analysis. First was the time and effort it took to develop WoT and program the integration between WoT and the CG. The second cost was the large amount of computational time it took to run WoT and later debug the integration between WoT and the CG. These delays significantly slowed down the iterations.

An analytical approach, similar to Loeb's, might have estimated sufficiency of different portfolios of weapons using relatively simple formulas that considered Pks, target inventories, and weapons portfolios, while sacrificing some fidelity. Such an approach might have been easier to integrate with the CG and might have taken much less time to compute.

Table B.5
Example Form of Results Database Used in the Analysis

Xs					Ls			Ms	
0.38	0.14	0.01	0.54	0.75	0.14	0.01	0.68	0.34	0.28
0.56	0.72	0.29	0.07	0.34	0.90	0.52	0.54	0.66	0.51
0.27	0.19	0.51	0.46	0.59	0.79	0.05	0.92	0.04	0.97
0.65	0.57	0.86	0.75	0.67	0.91	0.50	0.68	0.45	0.44
0.31	0.60	0.93	0.23	0.61	0.19	0.46	0.22	0.49	0.31
0.00	0.93	1.00	0.20	0.95	0.11	0.36	0.27	0.21	0.62
0.67	0.38	0.67	0.99	0.28	0.11	0.18	0.39	0.25	0.23
0.92	0.39	0.33	0.72	0.41	0.53	0.98	0.81	0.08	0.13
0.86	0.48	0.40	0.63	0.32	0.01	0.11	0.39	0.32	0.99
0.65	0.40	0.82	0.22	0.80	0.67	0.55	0.48	0.39	0.02
0.36	0.03	0.30	0.11	0.74	0.85	0.21	0.02	0.39	0.68
0.55	0.11	0.42	0.61	0.50	0.81	0.87	0.84	0.17	0.02
0.33	0.61	0.00	0.09	0.07	0.78	0.15	0.04	0.89	0.43
0.08	0.85	0.41	0.65	0.28	0.75	0.89	0.37	0.27	0.19
0.99	0.45	0.73	0.91	0.36	0.05	0.38	0.61	0.08	0.90
0.20	0.51	0.16	0.15	0.69	0.78	0.39	0.49	0.46	0.41

Data

This appendix details the data that we used to populate the CG (in Analytica), which is used to construct campaign datasets that are simulated within the WoT model.

For the most part, the data we used are based on educated guesses, research on open sources, and simplifying assumptions, since the real data are unavailable, classified, or more complicated.

For easy reference, this appendix is ordered in the same way as data in the WoT input file.

Security Environments, Security Conditions, and Campaigns

Each two-year period of the model is defined by a security condition. Table C.1 shows the names of the six security conditions. Each security condition lasts a minimum length of time, between one and three two-year periods.

During each security condition, we assume that a mix of campaigns occurs. Table C.2 shows the relationship between each two-year period of a security condition and the campaigns conducted. The CG generates campaign data for each campaign to pass to the WoT model. If there are multiple campaigns within a security condition, the CG assumes that the campaigns occur in succession, and the munitions depleted in one campaign are removed from inventories for later campaigns.

Security environments are a time line of ten two-year security conditions. There is a very large number of possible security environments; the security environments that occur are an uncertainty. In earlier stages of the project, we developed nine representative security environments. We supplemented these with 11 additional security environments in the first phase of

Table C.1
Security Conditions and Minimum Durations

Security Condition		Duration
A	Quiescent	1
B	Deterrent	3
C	Long major regional	2
D	Major regional campaign	2
E	Multiple major regional	1
F	Global war	2

Table C.2
Campaigns Conducted Within Each Two-Year Period of a Security Condition

Security Condition		Incident	Global Incident	Deterrence	Short Campaign	Major Regional and COIN	Major Regional	Massive Regional
A	Quiescent	1	1	0	0	0	0	0
B	Deterrent	0	0	1	0	0	0	0
C	Long major regional	0	0	0	0	1	0	0
D	Major regional campaign	0	0	0	1	0	1	0
E	Multiple major regional	0	0	0	0	0	2	0
F	Global war	0	0	0	0	0	1	2

the analysis, which attempt to maximize the variability of all of the security environment in terms of their magnitude, variability, and trend. In the second phase, we added a security environment that approximated the past 20 years (SE 21) and an additional four randomly generated SEs with lower severity. The resulting 25 security environments are shown in Figure 2.3.

Permissiveness and Nonpermissiveness

In a permissive environment, all weapons can be used. In a nonpermissive environment, only certain weapons can be used (see weapons attributes, below). We assume that all of the campaigns except major regional and COIN begin in a nonpermissive environment. To achieve a permissive environment, all targets at or above the cutoffs in Table C.3 need to be destroyed. (See also Table C.8, below.)

Table C.3
Permissiveness Target Priority Cutoffs

Campaign	Cutoff
Incident	5
Global incident	7
Deterrence	14
Short campaign	5
Major regional and COIN	N/A
Major regional	5
Massive regional	7
Global war	14

This method of determining permissiveness is relatively simplistic. Future development efforts might want to differentiate specific, air-defense targets that create a nonpermissive environment. Such an effort is not trivial because it would require a relaxation of the implicit assumption that the highest-priority targets are also the targets that create a nonpermissive environment.

Maximum Number of Days for Campaigns

We assume that all campaigns must be completed in 200 days to be successful. If campaigns last longer than 200 days, they are considered to have failed, the campaign stops, and no additional weapons are expended.

Future extensions may modify the maximum number of days to be different for each type of campaign. For example, it is likely that 200 days would be politically unacceptable for an incident, but more than 200 days might be necessary for a massive regional campaign.

Weapons Groups and Weapons Delivery Rates

WoT assumes that each weapon belongs to a single weapons group. Each weapons group has a maximum delivery rate (both a maximum surge delivery rate and a maximum steady-state delivery rate). Table C.4 shows the maximum weapons delivery rates for each of the four weapons groups. All of the weapons groups are assumed to have seven days of surge, except for the Naval weapons group, whose surge rate and steady-state rates are the same. These data are highly uncertain, which is why we allow ± 50 percent uncertainty in these rates.

WoT's treatment of weapons delivery is currently highly simplified from reality, where weapons can be delivered from many different platforms. Future extensions may allow this many-to-many relationship (instead of the current many-to-one relationship) and model each weapons group as a single platform. Such an extension would be much more computationally complex than the present model.

Weapons delivery rates are customized for each type of campaign to be some percentage of the maximum delivery rate based on the severity of the campaign. Table C.5 assumes that more stressful campaigns are provided with more delivery platforms.

Table C.4
Maximum Weapons Delivery Rates

Group	Group Name	Surge Days	Surge Rate (Maximum Daily)	Steady-State Rate (Maximum Daily)
0	Nonstealth	7	2,500	1,500
1	Stealth	7	500	300
2	Naval	999	100	100
3	Land	7	50	40

Table C.5
Adjustments to Weapons Delivery Rate for Each Campaign Type

Campaign	Adjustment (% of Maximum Daily)
Massive regional	100
Major regional	90
Major regional and COIN	80
Short campaign	70
Deterrence	60
Global incident	50
Incident	40

Munitions Attributes

We selected 30 types of weapons that are currently in stockpiles of U.S. conventional weapons. Table C.6 lists these 30 weapons and their attributes. Each weapon is assigned to one weapons group (as detailed above).

Designation as a PGM weapon or a GPS weapon is important for applying the weapons effectiveness uncertainties. Weapons that use lasers, radars, or video for seekers are designated as PGMs. Weapons that use GPS for navigation and targeting are designated as GPS weapons. A future extension may also incorporate inertial and terrain navigation, which is incorporated into some types of missiles and may make them less susceptible to degradations in GPS.

The models assume that weapons cannot be used in nonpermissive conditions unless they are designated as nonpermissive weapons. Eleven of the 30 weapons are designated as nonpermissive.

Portfolio baselines and unit costs were found by looking at estimates in open sources or by making our own estimates. There are likely to be uncertainties in these numbers, especially with the unit costs. For example, many reported unit costs appear to be average costs for weapons programs. For ongoing weapons procurements, marginal costs are likely to be lower than average costs. However, unit costs for older weapons might require restarting production lines, which could increase the unit costs. A future extension could incorporate uncertainty into weapons costs; however, the key uncertainties are relative costs, which would be more difficult to incorporate because of the large number of weapons.

Days Between Replenishments

WoT has the ability to replenish weapons at specified time intervals. At present, the model assumes that there are no replenishments.

Two future extensions may necessitate the use of the replenishment feature. First, logistics could be incorporated into the models, which would account for the fact that munitions must be moved into theater, so the munitions available to a campaign may grow over time. Such an

Table C.6
Weapons Used in the Models and Their Attributes

ID	Weapon	Weapons Group	PGM	GPS	Nonpermissive Weapon?	Unit Cost ($ Thousands)
W00	AGM-130C (BLU 109)	Nonstealth		Yes	Yes	450
W01	AGM-130A (Mk 84)	Nonstealth		Yes	Yes	450
W02	AGM-154 (JSOW)	Nonstealth		Yes	Yes	719
W03	AGM-158 (JASSM-ER)	Nonstealth		Yes	Yes	1512
W04	AGM-64 (Maverick)	Nonstealth	Yes			158
W05	AGM-84H (SLAM/ER)	Nonstealth		Yes	Yes	1,200
W06	AGM-86 (CALCM)	Nonstealth		Yes	Yes	1,160
W07	AGM-88E (HARM/AARGM)	Stealth	Yes	Yes	Yes	200
W08	BGM-109E (TLAM)	Naval		Yes	Yes	1,576
W09	GBU-10 Paveway II (BLU 109)	Nonstealth	Yes			24
W10	GBU-10 Paveway II (Mk 84)	Nonstealth	Yes			24
W11	GBU-12 (Paveway II)	Nonstealth	Yes			19
W12	GBU-15 (BLU-109)	Nonstealth	Yes			28
W13	GBU-15 (Mk 84)	Nonstealth	Yes			28
W14	GBU-24 Paveway III (BLU-109)	Nonstealth	Yes			55
W15	GBU-24 Paveway III (BLU-116)	Nonstealth	Yes			55
W16	GBU-24 Paveway III (Mk 84)	Nonstealth	Yes			55
W17	GBU-28 Bunker Buster	Stealth	Yes			145
W18	GBU-31 JDAM (2,000 lb)	Stealth		Yes		62
W19	GBU-32 JDAM (1,000 lb)	Stealth		Yes		62
W20	GBU-36/37 (GAM)	Stealth		Yes	Yes	231
W21	GBU-38 JDAM (500 lb)	Stealth		Yes		62
W22	GBU-39 SDB	Nonstealth		Yes		29
W23	GBU-53B SDB II	Nonstealth	Yes	Yes		40
W24	GBU-54 (Laser JDAM)	Stealth	Yes	Yes		21
W25	GBU-57B MOP	Stealth		Yes	Yes	4,000
W26	MGM-168 ATACMS	Land		Yes	Yes	820
W27	Mk-82	Nonstealth				5
W28	Mk-83	Nonstealth				10
W29	Mk-84	Nonstealth				15

extension would require some additional modifications to the models to ensure that weapons use never exceeded global munitions inventories. The second type of extension is weapons production during a campaign. Currently, the CG assumes that weapons will be produced instantly when acquisition funding is spent. In reality, some weapons will be delivered during a campaign. Although a 200-day campaign is probably too short to order and produce a significant number of new weapons (with a possible exception of dumb bombs, which we assume are virtually unlimited in number because of their high baseline numbers), history has shown that new weapons can be designed, tested, and produced in a short amount of time.[1]

Target Attributes

Thirty types of targets were chosen for these models. The 30 were chosen for being representative of the diversity of targets that may be found in actual conflicts. Table C.7 lists the targets and their attributes. Most of the attributes in Table C.7 are not used directly in the models but were used as guides for choosing weapons with a diverse set of attributes and for developing estimates of Pk values.

The first two target attributes are mobility, which is differentiated between mobile (Mb) targets and fixed (Fx) targets (if the target's aimpoint is a latitude/longitude), and hardness, which is differentiated between hard (Hd) targets requiring a blast warhead, soft (Sf) targets better suited for fragmenting warheads, and buried (Br) targets requiring penetrating warheads. This gives six possible classes of targets.

Another attribute that was created but was not used is the permissiveness attribute, which designates whether a target can be targeted in a nonpermissive environment. Instead, WoT assumes that all targets can be targeted in nonpermissive environments. The addition of permissive (Pm) targets and nonpermissive (Np) targets gives 12 possible classes of targets.

The dispersion attribute addresses the number of target aimpoints: This differentiates between single point (Pt) targets and area (At) targets. The dispersion attribute is based on whether the target can be serviced with a single weapon or needs multiple weapons to cover it. In other words, a target requiring four weapons would need to have four individual successful hits to kill the target. Area targets might come in a variety of sizes (e.g., four weapons needed for a barracks, eight weapons needed for a factory, 16 weapons for an airfield, and 32 for a supply depot). Therefore, there are four classes of area targets (At4, At8, At16, and At32, respectively), the addition of which gives 60 possible classes of targets. An extension to the WoT model is being developed that accounts for dispersed targets.

The collateral damage attribute addresses the target's proximity to entities that are protected by the laws of international armed conflict or political considerations—places that it would be counterproductive to hit. This attribute differentiates between sensitive (Cs) targets and nonsensitive (Ns) targets. The collateral damage attribute is based on the size of the weapon that can be used on the target. For targets in a congested urban environment, only 250 lb warheads might be allowed, otherwise the Pk would be 0 for larger warheads. Similarly, some targets would have upper limits of 500 lb or 1,000 lb (2,000 lb would, by definition, be

---

[1] For example, the GBU-28 Bunker Buster was initially conceived, designed, built, tested, and used following the onset of Operation Desert Storm when existing weapons were revealed to be insufficient for targeting Iraq's underground bunkers.

Table C.7
Classes of Targets and Their Attributes

Domain	Real World Targets	Mobility (Mb or Fx)	Hardness (Sf, Hd, or Br)	Permissive-ness (Pm or Np)	Dispersion (Pt, At4, At8, At16, or At32)	Collateral Damage (Cs250, Cs500, Cs1000, or Ns)	Time Criticality (Ft or It)
Air	Aircraft on the ground	Mb	Sf	Np	At16	Ns	Ft50%
Air	Aircraft shelters	Fx	Br	Np	At16	Ns	It
Air	Airfields	Fx	Hd	Np	At16	Ns	It
Ground	Barracks	Fx	Hd	Pm	At4	Cs250	It
Ground	Bunkers	Fx	Br	Np	Pt	Ns	It
C2	C2 headquarters	Fx	Br	Np	Pt	Cs1000	It
Comm	Comms RF systems	Fx	Sf	Np	Pt	Cs250	It
Comm	Comms satellite downlinks	Fx	Sf	Np	Pt	Cs1000	Ft50%
Comm	Communication equipment	Fx	Sf	Np	Pt	Cs250	It
Comm	Communication facilities	Fx	Hd	Np	At4	Cs250	It
Logistics	Depots	Fx	Hd	Np	At32	Ns	It
Comm	Fiber optic systems	Fx	Br	Np	At8	Ns	It
C2	Intelligence processing center	Fx	Hd	Np	At4	Cs500	It
Sensor	Land-based passive detection	Mb	Sf	Pm	At4	Ns	It
Sensor	Land-based radar	Fx	Sf	Pm	At16	CS1000	It
Ground	Personnel carrier formation	Mb	Sf	Pm	At32	Ns	Ft05%
Logistics	POL refinery	Fx	Hd	Np	At4	Ns	It
Naval	Port infrastructure	Fx	Hd	Np	At8	Ns	It
Naval	Port systems	Mb	Sf	Np	At4	Ns	It
Naval	Ships at sea	Mb	Hd	Np	At32	Ns	Ft50%
Naval	Ships in port	Mb	Hd	Np	At4	Ns	It
Naval	Submarine pens	Fx	Br	Np	At4	Ns	It
Ground	Surface-to-air missiles	Mb	Sf	Np	Pt	Ns	Ft20%
Ground	Surface-to-surface missiles	Mb	Sf	Pm	Pt	Ns	Ft01%
Ground	Tank—individual	Mb	Hd	Pm	Pt	Ns	Ft20%
C2	Terror cell meeting place	Mb	Sf	Pm	Pt	Cs250	Ft05%
Ground	Troops in the field	Mb	Sf	Pm	At32	Cs500	Ft20%
Logistics	Weapon transport—systems	Mb	Sf	Pm	At8	Ns	Ft50%
Logistics	Weapon transport infrastructure	Fx	Hd	Pm	Pt	Ns	It
Logistics	WMD storage sites	Fx	Br	Pm	Pt	Ns	Ft05%

NOTES: Mb = mobile; Fx = fixed; Sf = soft; Hd = hard; Br = buried; Pm = permissive; Np = nonpermissive; Pt = point; At = area targets (not implemented in this study); Ns = nonsensitive collateral damage; Cs = sensitive collateral damage; It = invariant targets; Ft = fleeting targets (not implemented in this study).

an Ns target). Therefore, there could be up to three sensitive target attributes (Cs250, Cs500, Cs1000), the addition of which gives 240 possible classes of targets.

The time-criticality attribute addresses the target's temporal availability; this differentiates between fleeting (Ft) targets and invariant (It) targets. The time-criticality attribute is based on how long the target would be available for striking. The time-criticality attribute is based on whether the target is available when the launch platform is at its launch point, resulting in a binary 1 or 0 Pk, similar to permissiveness. Each fleeting target is distinguished by the percentage of days the target is revealed and can be targeted. There are four possible percentages, Ft50, Ft20, Ft05, and FT01, the addition of which gives 1,200 possible classes of targets. We are currently developing an extension to WoT that incorporates fleeting targets, but it has not yet been tested sufficiently to include it in the runs reported here. Initial testing suggests that fleeting targets will extend the length of campaigns substantially—it may be difficult to destroy an entire class of targets that is exposed 1 percent of the time in 200 days.

To choose 30 target classes from the 1,200 possible target classes, we selected representative targets and used guidance produced by the Chief of Naval Operations for Air-Sea Battle (Greenert, 2012).

Each campaign type has a unique subset of included targets with a unique distribution and prioritization of those targets. Tables C.8a and 8b show this information for all 30 target classes.

Each target is assigned a priority, between 1 and the number of targets within the campaign. When WoT decides which targets to strike in each period, it starts with the highest-priority target (e.g., a target with a priority of 1) and works its way down the priority list until all targets have been considered for targeting or until all weapons have been expended or all delivery vehicles have been committed.

The shading of target priorities indicates whether a target is above or below the permissiveness cutoff for that campaign. If targets remain with priorities shaded in white, they cause the campaign to remain in a nonpermissive state. Targets with priorities shaded in green do not affect the permissive state of the campaign.

Target prioritization relied on reasoning from team members and the Air-Sea Battle guidance produced by the Chief of Naval Operations that was used to select the 30 target classes (Greenert, 2012). Priorities in the incident/minor campaign were set so that WoT first suppresses enemy air defenses then targets command and control (C2), then WMD capability, and finally warfighting capability. In global incident, WoT first eliminates the WMD threat, then targets near-term threatening terrorist nodes, then the leadership cadre, then communications, and finally foot soldiers. The Deterrence campaign is a rehearsal for a massive regional campaign, except with a higher prioritization of decapitation. The short campaign has a similar prioritization to the incident campaign but with a slightly higher emphasis on aircraft on the ground. Major regional and COIN has the same prioritization as the global incident, except it includes some additional conventional forces. Priorities in major regional are designed to represent total, conventional war, except some elements of adversary infrastructure and industrial capability are left intact to lessen the reconstruction burden. Finally massive regional is a similar total, conventional war with more targeting of adversary infrastructure; however, some leadership is left intact to avoid precipitating the campaign to nuclear.

The second column for each campaign in Tables C.8a and C.8b shows the share of the total targets that each target type represents in the campaign. The total number of targets for each campaign is shown at the bottom of the table. Incident/minor campaign and global

Table C.8a
Priorities and Distribution of Targets Within Campaigns

Domain	Real World Targets	Incident/Minor Campaign (Desert Fox/Somalia/Tanker War)		Global Incident (GWoT-Like)		Deterrence (Cold War)		Short Campaign (Grenada/Panama/Kosovo/Libya)	
		Priority	%	Priority	%	Priority	%	Priority	%
Air	Aircraft on the ground	15	3			16	2	14	3
Air	Aircraft shelters					20	3		
Air	Airfields	9	8			11	2	9	10
Ground	Barracks	11	4	6	20	25	2	11	3
Ground	Bunkers			5	10	27	1		
C2	C2 headquarters	6	5	3	10	1	15	6	5
Comm	Comms RF systems	3	3			8	1	3	3
Comm	Comms satellite downlinks					9	4		
Comm	Communication equipment	12	3			6	2	12	3
Comm	Communication facilities	7	5	4	5	7	1	7	4
Logistics	Depots					28	3		
Comm	Fiber optic systems	4	3			10	3	4	5
C2	Intelligence processing center	13	5			2	7	13	5
Sensor	Land-based passive detection	1	3			5	1	1	3
Sensor	Land-based radar	2	15			4	3	2	12
Ground	Personnel carrier formation					18	3		
Logistics	POL refinery					29	3		
Naval	Port infrastructure					26	1		
Naval	Port systems					22	2		
Naval	Ships at sea					23	5		
Naval	Ships in port					21	2		
Naval	Submarine pens					12	2		
Ground	Surface-to-air missiles	5	15			14	7	5	16
Ground	Surface-to-surface missiles	8	10			15	6	8	13
Ground	Tank—individual					19	5		
C2	Terror cell meeting place			2	35				
Ground	Troops in the field			7	15	24	2		
Logistics	Weapon transport—systems					17	5		
Logistics	Weapon transport infrastructure	14	3			13	2	15	5
Logistics	WMD storage sites	10	15	1	5	3	5	10	10
Target totals for each campaign type		340		750		1,500		3,000	

Table C.8b
Priorities and Distribution of Targets Within Campaigns

Domain	Real World Targets	Major Regional and COIN (Afghanistan)		Major Regional (Korea/Vietnam/Desert Storm/Operation Iraqi Freedom)		Massive Regional (WWI)	
		Priority	%	Priority	%	Priority	%
Air	Aircraft on the ground			18	3	22	3
Air	Aircraft shelters			23	3	11	2
Air	Airfields			9	5	10	3
Ground	Barracks	6	20	11	3	12	2
Ground	Bunkers	5	10			27	5
C2	C2 headquarters	3	10	6	2	16	2
Comm	Comms RF systems	10	2	3	2	15	3
Comm	Comms satellite downlinks			14	2	1	1
Comm	Communication equipment	9	5	12	1	14	2
Comm	Communication facilities	4	5	7	1	4	2
Logistics	Depots			24	3	28	2
Comm	Fiber optic systems			4	1	5	1
C2	Intelligence processing center	8	5	13	2	6	2
Sensor	Land-based passive detection			1	1	2	1
Sensor	Land-based radar			2	2	3	3
Ground	Personnel carrier formation	11	3	21	5	25	5
Logistics	POL refinery					29	1
Naval	Port infrastructure					26	1
Naval	Port systems					19	3
Naval	Ships at sea			17	5	21	5
Naval	Ships in port			15	3	18	3
Naval	Submarine pens					17	4
Ground	Surface-to-air missiles			5	10	7	10
Ground	Surface-to-surface missiles			8	15	9	10
Ground	Tank—individual			22	5	13	3
C2	Terror cell meeting place	2	20				
Ground	Troops in the field	7	15	19	15	23	10
Logistics	Weapon transport—systems			20	3	24	3
Logistics	Weapon transport infrastructure			16	5	20	3
Logistics	WMD storage sites	1	5	10	3	8	5
Target totals for each campaign type		6,000		20,000		60,000	

incident each have a relatively small number of targets. Deterrence has more targets, but the emphasis of deterrence is provide a show of force across the variety of target types that might be found in much greater numbers in a major regional campaign. Short campaign has 3,000 targets, which is similar in magnitude to the campaign in Kosovo. Major regional and COIN has twice as many targets (6,000). Major regional has 20,000 targets, which was designed to be about two-thirds the size of Operations Desert Storm and Iraqi Freedom. Finally, massive regional was chosen to have three times as many targets (60,000) as major regional.

Pk Table

The Pk tables (Tables C.9a, b, and c) were constructed to show the probability of kill when matching any of the 30 weapons to any of the 30 target types.

The first step in creating the Pk table was setting Pk values to 0 for any combination of weapon and target that was inappropriate based on the attributes of the weapons and targets. For example, Tables C.9a and C.9b say that collateral damage is an issue when targeting barracks, so no weapon with a warhead bigger than 250 lb (CS250) may be used; therefore, all Pks for weapons with a warhead over 250 lb are 0 for the Barracks row.

The second step in creating the Pk was setting nonzero Pk values. Theses Pk values were chosen by making educated guesses, taking into account the attributes of each weapon and target. These guesses were made in part by setting relative Pk values taking into account expectations about the relative effectiveness of weapons, for example, by different weapons across a target class or different targets for a type of weapon. A future extension is to use classified Pk values, which should be more accurate than these estimates.

Defense/Munitions Funding Levels

Unlike the previous data in this appendix, data used for munitions funding are used entirely within the CG, which selects the munitions that it purchases based on the strategy (e.g., the portfolio goals) and the funding available to purchase munitions.

To simulate defense funding levels, the team looked at historic defense funding levels. Figure C.1 shows that the pattern of defense funding as a share of the GDP has historically spiked depending on the severity of conflicts.

Munitions funding is assumed to be capped at 0.85 percent of the DoD budget, as confirmed through CAPE. The DoD budget is assumed to be set for each two-year period depending upon the security environment in the previous two-year period, as shown in Table C.10. Therefore, there will be a lag in funding.[2]

Gross domestic product (GDP) is expressed in fiscal year (FY) 2013 dollars (i.e., in the same units as the estimated costs of munitions) and is assumed to grow steadily as shown in Table C.11.

---

[2]   Because the initial period of each security environment is not preceded by another security environment, we assume that the share of GDP in the first period is 3.0 percent of GDP for all security environments.

Table C.9a
Pk Table for Weapons 0–9

Real World Targets	W00: AGM-130C (BLU 109)	W01: AGM-130A (Mk 84)	W02: AGM-154 (JSOW)	W03: AGM-158 (JASSM-ER)	W04: AGM-64 (Maverick)	W05: AGM-84H (SLAM/ER)	W06: AGM-86 (CALCM)	W07: AGM-88E (HARM/AARGM)	W08: BGM-109E (TLAM)	W09: GBU-10 Paveway II (BLU 109)
Aircraft on the ground	0.1	0.95	0.95	0.98	0.8	0	0	0	0.5	0.05
Aircraft shelters	0.95	0.9	0	0	0	0	0	0	0	0.3
Airfields	0.98	0.98	0.1	0.05	0.5	0	0.98	0	0.9	0.98
Barracks	0	0	0.4	0.4	0.2	0.4	0	0	0	0
Bunkers	0.9	0.95	0.5	0.95	0	0.4	0	0	0	0.3
C2 HQ/one-story building	0	0	0.4	0.4	0.2	0.4	0	0	0.5	0
Comms RF systems	0	0	0.4	0.4	0.2	0.4	0	0	0	0
Comms satellite downlinks	0	0	0.4	0.4	0.2	0.4	0	0	0	0
Communication equipment	0	0	0.4	0.4	0.2	0.4	0	0	0	0
Communication facilities	0	0	0.4	0.4	0.2	0.4	0	0	0	0
Depots	0.05	0.05	0	0.05	0	0	0	0	0	0.05
Fiber optic systems	0.7	0.95	0.4	0.4	0.2	0.4	0.6	0	0.5	0.6
Intelligence processing center	0	0	0.4	0.4	0.2	0.4	0	0	0	0
Land-based passive detection	0.9	0.95	0.4	0.4	0.2	0.4	0.6	0	0.5	0.6
Land-based radar	0	0	0.1	0.98	0.8	0	0	0.9	0.95	0
Personnel carrier formation	0.05	0.1	0.4	0.05	0.7	0	0	0	0	0
POL refineries	0.98	0.98	0.8	0.98	0.98	0	0.95	0	0.8	0.9
Port infrastructure	0.4	0.95	0.95	0.1	0.05	0.1	0.4	0	0	0.2
Port systems	0.1	0.1	0.2	0.05	0.8	0	0	0	0	0
Ships at sea	0.25	0.5	0.5	0.98	0.4	0.4	0	0	0	0
Ships in port	0.5	0.5	0.4	0.9	0.5	0.4	0	0	0.5	0
Submarine pens	0.9	0.1	0.5	0.95	0.2	0.4	0	0	0	0.4
Surface-to-air missiles	0.05	0.1	0.05	0.05	0.8	0	0	0	0	0.8
Surface-to-surface missiles	0.5	0.98	0.95	0.98	0.8	0	0	0	0	0
Tank—individual	0.05	0.05	0.3	0	0.8	0	0	0	0	0
Terror cell meeting place	0	0	0.95	0.4	0.2	0.4	0	0	0	0
Troops in the field	0	0	0.95	0.05	0	0	0	0	0	0
Weapon transport—systems	0.9	0.95	0.5	0.7	0.05	0	0	0	0	0.3
Weapon transport infrastructure	0.9	0.95	0.95	0.95	0.4	0.2	0	0	0	0
WMD storage sites	0	0	0	0	0	0	0	0	0	0

Table C.9b
Pk Table for Weapons 10–19

Real World Targets	W10: GBU-10 Paveway II (Mk 84)	W11: GBU-12 (Paveway II)	W12: GBU-15 (BLU-109)	W13: GBU-15 (Mk 84)	W14: GBU-24 Paveway III (BLU-109)	W15: GBU-24 Paveway III (BLU-116)	W16: GBU-24 Paveway III (Mk 84)	W17: GBU-28 Bunker Buster	W18: GBU-31 JDAM (2,000 lb)	W19: GBU-32 JDAM (1,000 lb)
Aircraft on the ground	0.8	0.8	0.95	0.95	0.5	0.95	0.95	0	0.98	0.98
Aircraft shelters	0.2	0	0	0.7	0.6	0	0.7	0.95	0.95	0.5
Airfields	0.98	0.7	0.98	0.05	0.05	0	0.05	0	0.05	0.98
Barracks	0	0	0	0	0	0	0	0	0	0
Bunkers	0.5	0	0.9	0.95	0	0	0.6	0	0.6	0.2
C2 HQ/one-story building	0	0.5	0	0	0	0	0	0	0	0.4
Comms RF systems	0	0	0	0	0	0.4	0	0	0	0
Comms satellite downlinks	0	0	0	0	0	0.4	0	0	0	0.4
Communication equipment	0	0	0	0	0	0.4	0	0	0	0
Communication facilities	0	0	0	0	0	0.4	0	0	0	0
Depots	0.05	0	0.05	0.05	0.05	0	0.05	0	0.05	0.05
Fiber optic systems	0.8	0.4	0.8	0.7	0	0	0.9	0	0.95	0.4
Intelligence processing center	0	0.4	0	0	0	0.4	0	0	0	0
Land-based passive detection	0.8	0.4	0.8	0.95	0.7	0.4	0	0	0.95	0.3
Land-based radar	0	0.8	0	0	0	0.95	0	0	0	0.1
Personnel carrier formation	0.5	0.5	0.95	0	0	0.95	0.05	0	0.95	0.8
POL refineries	0.9	0.95	0.98	0.98	0.98	0	0.98	0	0.98	0.98
Port infrastructure	0.7	0	0.1	0.95	0.4	0	0.8	0	0.9	0.4
Port systems	0.4	0.7	0	0	0	0	0.9	0	0.1	0.05
Ships at sea	0.4	0.1	0	0.6	0	0	0	0	0	0.1
Ships in port	0.4	0.1	0	0.9	0	0.95	0.5	0	0	0.1
Submarine pens	0.5	0.5	0.9	0.95	0.5	0	0.6	0	0.6	0.2
Surface-to-air missiles	0.8	0.8	0	0	0.9	0.95	0.95	0	0.05	0.05
Surface-to-surface missiles	0.8	0.8	0.95	0.95	0.5	0.95	0.95	0	0.98	0.98
Tank—individual	0.2	0	0	0	0	0	0.8	0	0.05	0
Terror cell meeting place	0	0	0	0	0	0	0	0	0	0
Troops in the field	0	0.4	0	0	0	0	0	0	0	0
Weapon transport—systems	0.4	0.05	0.4	0.98	0.05	0	0.8	0	0.9	0.8
Weapon transport infrastructure	0.2	0	0	0.9	0.5	0	0.4	0	0	0.1
WMD storage sites	0	0	0	0	0	0	0	0	0	0

Table C.9c
Pk Table for Weapons 20–29

Real World Targets	W20: GBU-36/37 (GAM)	W21: GBU-38 JDAM (500 lb)	W22: GBU-39 SDB	W23: GBU-53B SDB II	W24: GBU-54 (Laser JDAM)	W25: GBU-57B MOP	W26: MGM-168 ATACMS	W27: Mk-82	W28: Mk-83	W29: Mk-84
Aircraft on the ground	0.98	0.98	0.9	0.9	0.95	0	0.4	0.1	0.15	0
Aircraft shelters	0.6	0	0	0	0	0.98	0	0	0.2	0.4
Airfields	0.98	0.7	0	0	0.7	0	0	0.05	0.1	0.2
Barracks	0	0	0	0	0	0	0.8	0	0	0
Bunkers	0.3	0	0	0	0	0	0	0	0.1	0.25
C2 HQ/one-story building	0	0.4	0.2	0.2	0.4	0	0.2	0.1	0	0
Comms RF systems	0	0	0.1	0.1	0	0	0.2	0	0	0
Comms satellite downlinks	0	0.4	0.2	0.2	0.4	0	0.2	0.1	0.1	0
Communication equipment	0	0	0.1	0.1	0	0	0.2	0	0	0
Communication facilities	0	0	0.1	0.1	0	0	0.2	0	0	0
Depots	0.1	0.1	0	0	0	0	0	0	0.1	0.2
Fiber optic systems	0.4	0.4	0.2	0.2	0.9	0	0.2	0.1	0.2	0.2
Intelligence processing center	0	0.4	0.2	0.2	0.4	0	0	0	0	0
Land-based passive detection	0.4	0.4	0.2	0.2	0.4	0	0.4	0.1	0.15	0.2
Land-based radar	0	0.1	0	0	0	0	0.4	0.1	0	0
Personnel carrier formation	0.85	0.7	0	0	0.7	0	0.6	0	0	0
POL refineries	0.98	0.98	0.8	0.8	0.85	0	0.8	0.6	0.7	0.8
Port infrastructure	0.5	0.4	0.2	0.2	0.25	0	0	0.1	0.2	0.3
Port systems	0.1	0.1	0	0	0.1	0	0.1	0.1	0.15	0.3
Ships at sea	0.2	0.1	0	0	0	0	0	0	0	0
Ships in port	0.15	0.1	0	0	0	0	0	0	0	0
Submarine pens	0.4	0	0	0	0	0.98	0	0	0	0.1
Surface-to-air missiles	0.1	0.05	0.05	0.05	0.05	0	0.2	0	0	0
Surface-to-surface missiles	0.98	0.9	0.1	0.1	0.1	0	0.2	0	0	0
Tank—individual	0	0	0	0	0	0	0	0	0	0
Terror cell meeting place	0	0	0.5	0.5	0	0	0.9	0	0	0
Troops in the field	0	0.5	0	0	0	0	0.8	0	0	0
Weapon transport—systems	0.95	0	0.4	0.4	0.5	0	0	0.1	0.2	0
Weapon transport infrastructure	0.15	0	0	0	0	0	0	0	0.1	0.25
WMD storage sites	0	0	0	0	0	0.9	0	0	0.1	0.1

Figure C.1
Historic Defense Spending as a Share of GDP

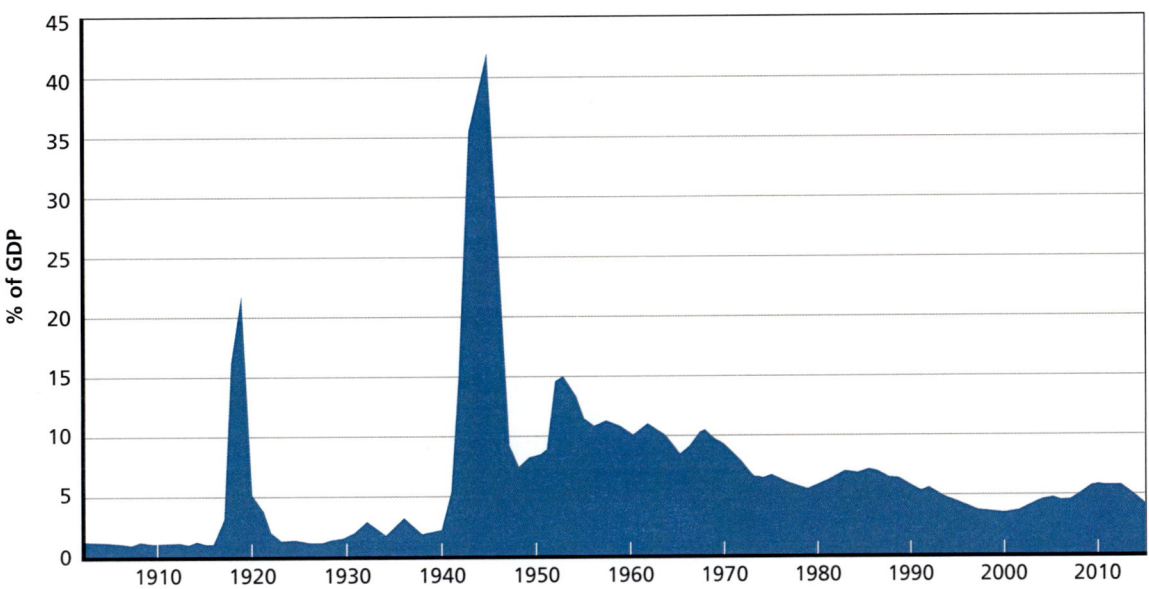

SOURCE: usgovernmentspending.com.
RAND *RR1112-C.1*

Table C.10
DoD Budget as a Percentage of GDP,
Based on the Security Environment
in the Previous Period

Security Environment	% of GDP
Major regional campaign	7.0
Quiescent	3.5
Deterrent	4.0
Long major regional	6.0
Multiple major regional	13.5
Global war	20.0

Table C.11
Growth of GDP
(Constant FY 2013 Dollars)

Year	GDP ($ Thousands)
2014	15.7T
2016	16.0T
2018	16.4T
2020	16.9T
2022	17.3T
2024	17.7T
2026	18.2T
2028	18.6T
2030	19.1T
2032	19.5T

References

Bankes, S. C. (1993). "Exploratory Modeling for Policy Analysis," *Operations Research*, **41**(3): 435–449.

Brooks, A., B. Bennett, and S. Bankes (1999). "An Application of Exploratory Analysis: The Weapon Mix Problem," *Military Operations Research*, **4**(1): 67–80.

Bryant, B. P. (2014). "sdtoolkit: Scenario Discovery Tools to Support Robust Decision Making." As of June 21, 2015:
http://cran.r-project.org/web/packages/sdtoolkit/index.html

Bryant, B. P., and R. J. Lempert (2010). "Thinking Inside the Box: A Participatory, Computer-Assisted Approach to Scenario Discovery," *Technological Forecasting and Social Change*, 77: 34–49.

Camm, F., L. Caston, A. C. Hou, F. E. Morgan, and A. J. Vick (2009). *Managing Risk in USAF Force Planning*, Santa Monica, Calif.: RAND Corporation, MG-827-AF. As of June 21, 2015:
http://www.rand.org/pubs/monographs/MG827.html

Danzig, R. (2011). *Driving in the Dark: Ten Propositions About Prediction and National Security*, Center for a New American Security, October 2011.

Davis, P. K. and Z. M. Khalilizad (1996). *A Composite Approach to Air Force Planning*, Santa Monica, Calif.: RAND Corporation, MR-787-AF. As of June 21, 2015:
http://www.rand.org/pubs/monograph_reports/MR787.html

Davis, P., R. D. Shaver, and J. Beck (2008). *Portfolio-Analysis Methods for Assessing Capability Options*, Santa Monica, Calif.: RAND Corporation, MG-662-OSD. As of June 21, 2015:
http://www.rand.org/pubs/monographs/MG662.html

Dewar, J. A. (2002). *Assumption-Based Planning: A Tool for Reducing Avoidable Surprises*, New York: Cambridge University Press.

DonVito, P. A. (1969). *The Essentials of a Planning-Programming-Budgeting System*, Santa Monica, Calif.: RAND Corporation, P-4124. As of June 26, 2015:
http://www.rand.org/pubs/papers/P4124

Fischbach, J. R. (2010). *Managing New Orleans Flood Risk in an Uncertain Future Using Non-Structural Risk Mitigation*, Santa Monica, Calif.: RAND Corporation, RGSD-262. As of June 21, 2015:
http://www.rand.org/pubs/rgs_dissertations/RGSD262.html

Greenert, Jonathan, Admiral (2012). "Projecting Power, Assuring Access," Chief of Naval Operations, May 10. As of June 23, 2015:
http://cno.navylive.dodlive.mil/2012/05/10/projecting-power-assuring-access/

Groves, D. G., M. Davis, R. Wilkinson and R. Lempert (2008). "Planning for Climate Change in the Inland Empire: Southern California," *Water Resources IMPACT*, July.

Groves, D. G., and R. J. Lempert (2007). "A New Analytic Method for Finding Policy-Relevant Scenarios," *Global Environmental Change*, 17: 73–85.

Hallegatte, S., A. Shah, R. Lempert, C. Brown and S. Gill (2012). *Investment Decision Making Under Deep Uncertainty: Application to Climate Change*, Washington, D.C.: World Bank.

Lempert, R. J., and S. W. Popper (2005). "High-Performance Government in an Uncertain World," in R. Klitgaard and P. Light (eds.), *High Performance Government: Structure, Leadership, Incentives*, Santa Monica, Calif., RAND Corporation, MG-256-PRGS. As of June 21, 2015: http://www.rand.org/pubs/monographs/MG256.html

Lempert, R. J., and M. Collins (2007). "Managing the Risk of Uncertain Threshold Responses: Comparison of Robust, Optimum, and Precautionary Approaches," *Risk Analysis,* **27**(4): 1009–1026.

Lempert, R. J., S. W. Popper, and S. C. Bankes (2003). *Shaping the Next One Hundred Years: New Methods for Quantitative, Long-Term Policy Analysis*, Santa Monica, Calif.: RAND Corporation, MR-1626. As of June 21, 2015: http://www.rand.org/pubs/monograph_reports/MR1626.html

Lempert, R. J., D. G. Groves, S. W. Popper, and S. C. Bankes (2006). "A General, Analytic Method for Generating Robust Strategies and Narrative Scenarios," *Management Science,* **52**(4): 514–528.

Lempert, R. J., N. Kalra, S. Peyraud, Z. Mao, S. B. Tan, D. Cira and A. Lotsch (2013a). *Ensuring Robust Flood Risk Management in Ho Chi Minh City: A Robust Decision Making Demonstration*, Washington, D.C.: World Bank.

Lempert, Robert J., Steven W. Popper, David G. Groves, Nidhi Kalra, Jordan R. Fischbach, Steven C. Bankes, Benjamin P. Bryant, Myles T. Collins, Klaus Keller, Andrew Hackbarth, Lloyd Dixon, Tom LaTourrette, Robert T. Reville, Jim W. Hall, Christophe Mijere, and David J. McInerney (2013b). *Making Good Decisions Without Predictions: Robust Decision Making for Planning Under Deep Uncertainty.* Santa Monica, Calif.: RAND Corporation, RB9701. As of September 22, 2015: http://www.rand.org/pubs/research_briefs/RB9701.html

Loeb, S. (2005). *Zeroing In: A Capabilities-Based Alternative to Precision Guided Munitions Planning*, Santa Monica, Calif.: RAND Corporation, RGSD-195. As of June 25, 2015: http://www.rand.org/pubs/rgs_dissertations/RGSD195.html

Means, E., M. Laugier, J. Daw, L. Kaatz, and M. Waage (2010). *Decision Support Planning Methods: Incorporating Climate Change into Water Planning*, Water Utility Climate Alliance: 76.

Simon, H. A. (1956). "Rational Choice and the Structure of the Environment," *Psychological Review,* **63**(2): 129–138.

Tetlock, P. E. (2006). *Expert Political Judgment: How Good Is It? How Can We Know?* Princeton, N.J.: Princeton University Press.

U.S. Department of Defense (2006). *Quadrennial Defense Review Report,* Washington, D.C., February 6. As of February 10, 2014: http://www.defense.gov/qdr/report/Report20060203.pdf

———— (2010). *Quadrennial Defense Review Report,* Washington, D.C., February 1. As of February 10, 2014: http://www.defense.gov/qdr/QDR%20as%20of%2029JAN10%201600.pdf

———— (2013), *Defense Budget Priorities and Choices, Fiscal Year 2014,* Washington, D.C., April. As of February 10, 2014: http://www.defense.gov/pubs/DefenseBudgetPrioritiesChoicesFiscalYear2014.pdf

———— (2014). *Quadrennial Defense Review Report,* Washington, D.C., March 4. As of March 18, 2014: http://www.defense.gov/pubs/2014_Quadrennial_Defense_Review.pdf

U.S. National Research Council (2009). *Informing Decisions in a Changing Climate,* T. N. A. Press. Washington, D.C.: Panel on Strategies and Methods for Climate-Related Decision Support, Committee on the Human Dimensions of Climate Change, Division of Behavioral and Social Sciences and Education.

University of Michigan (undated). "Random Number Generation; Biostatistics, 615/815, Lecture 14. As of July 5, 2013: http://www.sph.umich.edu/csg/abecasis/class/2006/615.14.pdf